AQUACULTURE

Dr. Keshav Kumar Jha was born on 19th August 1966 at Kakwara, Banka (Bihar). He had his higher education at T. M. Bhagalpur University, Bhagalpur which conferred on him the degree of M. Sc. (Zoology) and Doctor of Philosophy in Science in 1996. He has specialized in Ichthyology and Fisheries. Besides he has obtained his B.Ed. degree from Annamalai University, Tamil Nadu and M. Ed. degree from Himachal Pradesh University, Shimla. He has attended many courses and training programmes organized by reputed National Institutions such as CIFE, CDRI, Bose Research Institute, NCERT, NNMRS, CIFRI, NBFGR, UGC etc. He has more than nineteen years of Teaching and Research experiences. He has worked as JRF/SRF of ICAR/CSIR. He is the member of different National and International Research organizations. He has published many research papers in National and International journals and has also participated and presented papers in National and International Symposiums, Seminars, Workshops, Conferences etc. He has been associated with research work funded by various agencies. Presently Dr. Jha is working as Head, Department of Zoology (FIST-DST Supported Department), Jawaharlal Nehru College, Pasighat (B++ NAAC Graded Institution), Arunachal Pradesh.

AQUACULTURE

Keshav Kumar Jha
M. Sc, M. Ed, Ph. D
Head
Department of Zoology,
Jawaharlal Nehru College,
Pasighat – 791 103
Arunachal Pradesh

2010
DAYA PUBLISHING HOUSE
Delhi - 110 035

© 2010 KESHAV KUMAR JHA (b. 1966–   )
ISBN 9788170359494

Published by   :   **Daya Publishing House**
                          A Division of
                          Astral International Pvt. Ltd.
                          – ISO 9001:2008 Certified Company –
                          4760-61/23, Ansari Road, Darya Ganj
                          New Delhi-110 002
                          Ph. 011-43549197, 23278134
                          E-mail: info@astralint.com
                          Website: www.astralint.com

Laser Typesetting   :   **Classic Computer Services**
                          Delhi - 110 035

Printed at   :   **Chawla Offset Printers**
                          Delhi - 110 052

PRINTED IN INDIA

— Dedicated to —

MY PARENTS

Smt. ARUNA JHA
and
Shri UGRA MOHAN JHA

Acknowledgements

I feel myself extremely fortunate to have come up with an idea in my mind to share my experiences and experiments based on the non-stop researches with the readers and learners of this trade courtesy the almighty WHO has always been an enlightening guiding force behind me.

As I think of those who stood by me and helped me complete this work, a lot of names come to my mind. I place on record my deep sense of gratitude to my reverent teacher Prof. N. C. Datta, Former Professor and Head, Department of Zoology, Calcutta University, Kolkata, who has very willingly taken the pains to see the contents and text of the book, has given the suggestions, comments and has written the foreword to the book. I am also grateful to my other reverent teachers, Prof. J. S. Dutta Munshi, F.N.A, F.N.A.Sc, Former Professor and Head, Department of Zoology, T. M. Bhagalpur University, Bhagalpur, Prof. R. S. Pandey, Former Professor, S. K. University, Dumka, and Prof. T. K. Ghosh, Department of Zoology, T. M. Bhagalpur University, Bhagalpur for their ideas, suggestions, comments and constant encouragement through the compilation of this work.

I also received the direct or indirect supports from Dr. S. Ayyapan, ADG, Fisheries, Government of India, Dr. Dilip Kumar, Director, CIFE, Mumbai; Dr. K. K. Vas, Director, CIFRI, Barrackpore, Kolkata; Director, CIFA, Bhubaneswar; Dr. W. S. Lakra, Director, NBFGR, Lucknow; Dr. P. Krishnaiah, Chief Executive, NFDB, Hydrabad; Dr. Krishna Mitra, Secretary, IFSI, Barrackpore, Dr. B. C. Jha, Principal Scientist, CIFRI, Barrakpore; Dr. M. Mukherjee, Dy. Director of Fisheries, Government of West Bengal; Sri Abhimanu Mishra, Former Scientist, BARC, Mumbai; Prof. M. M. Gosawmi, Department of Zoology, Guwahati University, Guwahati; Prof. S. P. Biswas, Head, Department of Life Sciences, Dibrugarh University, Dibrugarh; Prof. W. Vishwanath, Department of Life Sciences, Manipur University, Manipur; Dr. P. K. Ray, Department of Zoology, T. M. Bhagalpur University, Bhagalpur; Dr. P. K. Khan, Department of Zoology, Patna

University, Patna and Dr. D. N. Pandit, Department of Zoology, V. K. S University, Arrah.

I am thankful to Dr. M. Munawar, Research Scientist, Great Lakes Laboratory for Fisheries and Aquatic Sciences, Canada; Dr. B. R. Subba, Department of Zoology, Tribhuban University, Nepal; Dr. S. S. T Nasar, Former Research Scientist, IIRR, Philippines; Dr. B. K. Chakraborty, Fisheries Officer, Department of Fisheries, Government of Bangladesh for their suggestions and moral support.

I am also beholden to Dr. Joram Begi, Director of Higher and Technical Education, Government of Arunachal Pradesh; Sri Tayek Talom, Principal and Dr. S. K. Sinha, Department of English, Jawaharlal Nehru College, Pasighat for their constant encouragements.

I am also thankful to Sri Jintu Rajkowa for typing and DTP related work of this manuscripts. Thanks are also due to Sri Anil Mittal and Sri Sunil Mittal of M/S Daya Publishing House, New Delhi for publishing the book.

Last but not the least, I am very grateful to my parents for their blessings, my wife Mamta who spare me from domestic work to complete this job in time and my sons Aman Tushar Jha and Abishek Kumar Jha for helping in every way as per their own capacities. Needless to say that in a book of this length mistakes may occur and I would be thankful to my readers if they write to me pointing out the same without any hitch and hesitation whatsoever.

Keshav Kumar Jha

Dr. N.C Datta *M. Sc., Ph.D.*
F.Z.S.I., F.N.I.E., F.N.E.S.A., F.N.C., F.A. Sc.T., F.M.B.A
Former Professor & Head of the Department of Zoology
Former Coordinator, M. Phil in Environmental Science
University of Calcutta

Foreword

The remarkable development of agriculture in India has been possible by the successful practice of the strategy of *"Green Revolution"* which could produce and provide the staple food (carbohydrate) to the hungry millions of India. But Indian people as a whole are now suffering badly from *protein malnutrition* which has become a major menace. Besides animal husbandry, *aquaculture has the potentiality to produce quality animal protein.* Aquaculture has been defined as the science of farming of fin fish and shell fish as well as some useful aquatic plants. In India aquaculture is now developing as a promising industry which besides producing protein, will generate income, provide employment to the rural people, earn foreign exchange and will also help to alleviate poverty.

In our education system, aquaculture has been introduced as an *academic* as well as *vocational subject.* Therefore, the *need of the hour is to have a good book on aquaculture.* I am indeed very happy to see that Dr. K.K. Jha, an eminent and experienced teacher, has ventured to write such a book. *It is also gratifying to note that Dr. Jha has included all the important topics of the subject which will be discussed in his book.* I hope that the *teachers, taughts, concerned government and non-government personnel, entrepreneurs and others will find the book useful.* I am also confident that a good book will help us to accomplish *"blue revolution"* or *"aquaplosion"* which is also expected to promote food security of the country.

I congratulate Dr. Jha for undertaking such a responsible job and I wish him a success.

N C Datta

Prof. N.C. Datta

110/20, B.T. Road, Kolkata – 700 108, W.B., Phone: (033) 2577-2050

Contents

Chapter 1
General Introduction

Fishing and hunting are the oldest profession of mankind ever since men began to search for food. Knowledge of the availability of fish in India is considered to be as old as three millennium B. C. During the excavations of Mohenjodaro and Harappa of the Indus-Valley civilizations (2500 B.C–1500 B.C), some indications of fish used as food have been found. Fish culture in the Indian Sub-continent is hundreds of years old. Kautilya's Arthashastra mentions that during war time, fishes in the reservoirs were rendered poisonous by secret means. This indicates that fish culture flourished at that time in reservoirs.

The importance of aquaculture in India or any other countries can be realized because the actual food production is not adequate to its actual demand. Moreover, the population is increasing in geometrical progression whereas the production is increasing in arithmetical progression resulting into food shortage. However, highly developed technology and scientific knowledge is being applied to enhance the production level. The population of India has increased enormously from 30 crore to about 115 crore within a short span of 64 years, showing an alarming shortage of food production. The magnitude of the problem can be reviewed from the fact that every minute 40 human beings are born. In every year we add a population of a state like Haryana or a country like Australia. In this case it is needless to say that the importance of aquaculture in present scenario becomes very significant as all the countries of the world are developing with all the available natural resources for the uplift of the people.

In a country like India where percapita consumption of meat and milk is small, aquaculture product has special importance as a supplement to ill-balanced cereal diets. In fact, protein deficiency is the world's most serious human nutritional problem today and perhaps 30–40 per cent of the world's population suffers from protein malnutrition. In India after 64 years of Independence, a large number of population is suffering from protein malnutrition problem resulting into several other ailments.

Proteins are the essential constituents of our diet which help in development of cells repairing, formation of hormones, enzymes etc. in our body. The chief sources to fulfill the requirement of protein *i.e.* the aquaculture, provides not only the protein but also other food requirements of our population. It will reduce the loss of huge money and manpower.

Aquaculture in recent times has assumed importance as a source of affordable animal protein for masses with a significant potential for supplementing livelihood in a rural economy. Now its contribution to the economic growth and nutrition security is well recognized nationally and globally. The fish production in the country has witnessed eight-fold growth from the 0.82 million tones in the 1950s to over 6.4 million metric tones at present in which inland fisheries including aquaculture contribute 3.4 million metric tones. This is due to contributing through technology support to achieve this target.

In the global scenario, China is one of the leading countries in fish production and India holds third position in the fish production and second in aquaculture in the world. India's fish production increased substantially and at the end of 10th plan period the total fish production was 6.4 million metric tones. There is a vast potential of water resources, proper utilization of water resources can further increase productivity. The average production in India is not even 1000 kg/ha. Our country has different water bodies' *viz.* Coastline of about 8129 kms, Exclusive Economic Zone 2.02 million sq. km, Continental shelf 0.506 million sq. km, Rivers and Canals 1,97,024 km, Reservoirs 3.15 million ha, Ponds and Tanks 2.35 million ha, Oxbow lakes and derelict waters 1.3 million ha, Bakishwaters 1.24 million ha and Estuaries 0.29 million ha.

Fisheries have been the source of economy besides providing food and employment since the dawn of human history. In real term the word "FISH" could be explained as:

F for "Food": Provides proteineous food for the well built of the human health. It also provides the vitamin A and D along with other minerals such as calcium, phosphorus, iron etc.

I for "Income": It generates employment.

S for "Sports": It provides a good sport for the Anglers.

H for "Hobby": Providing a good hobby to the Aquarium lovers etc.

Contribution of fisheries to the GDP of India is 1.07 per cent and 4.7 per cent of the agriculture GDP. The per capita fish availability is 9.0 kg. The annual Export earnings is Rs.7, 200 crore and this sector provide the employment of 1400 million. India has exploited only 10 per cent of its potential in aquaculture.

The aquaculture sector of the country is yet to take a great leap towards gearing its production to cater to the overall protein demand. Fish protein is the best animal protein for good health. Proper utilization of water resources will enable us to boost production so that the requirement of per capita animal protein can increase and more export can be possible. In fact the growth of fisheries/aquaculture in the country, at present is 6 per cent per year which is higher than any of other agriculture sub-sector.

Even though the total coastline in our country is so vast, the coastal catch has been declining due to lack of coastal management and proper policy of Government of India. In the south coast, there is a uniform ban on trawling during the period of breeding (April-August). But in the East coast where the conservation of fish stock is very much necessary, only the month of April and May has been earmarked for ban on fishing. As a result, indiscriminate fishing by deep sea fishing vessels and foreign trawlers has been causing depletion of fish stock in the east coast.

We should bear in mind, aquaculture sector is a big industry, which covers production, employment generation, and marketing, foreign exchange earning and strengthening of the economy of the country by raising GDP.

River associated ecosystem *viz.*, resources, lakes and wetlands offer immense potential for fish production, with appropriate specific ecological management. The four major aquatic habitats of rivers, reservoirs, beels and Chawers and estuaries that comprise the resource for aquaculture is open water, open access systems. The challenge is to address the interests of multi stockholders and ensure equitable availability of quality water for different enterprises. We need to focus our studies on water requirement and management protocols for suitable aquaculture, environment flows and reducing in rivers, environmental impact assessment, eutrophication, bioremediation and eco-restoration, river linking impact, conservation and fish pass designs, estuarine and riverine ecology with reference to important fish species such as Hilsa. Fishery resources evaluation on GIS for most and stock enhancement in reservoirs and in term of cage and pen culture.

Growth of aquaculture linked to healthy ecosystems in terms of biodiversity, clean water and fish for tomorrow, needs to be addressed in terms of appropriate management decisions and policy issues. New tools of data mining and eco-maudling, remote sensing, GIS and environment biotechnology will have to be employed for bringing out site-specific aquaculture enhancement protocols. Equally important are policy, as we are declining with common property resources. Community-based aquaculture management is the demand of the day and we need to look at mechanisms that would suit at varied situations across the country. The management decisions will be effective only if a holistic assessment of ecosystem, health and fish production, with active participation of stakeholders who control and manage the water resources and communities having direct involvement with the resources, is made. In this backdrop, research will have to be turned to develop better technology and management plans to face the impending challenge of more production, quality aquaculture product as well as to mitigate the ecosystem aberrations.

Aquaculture could be defined as maximization of production of aquatic organisms in confined water or in the captivity particularly in different aquatic habitat *viz.* freshwater, brackish water and seawater. The Indian fisheries/aquaculture sector has now been recognized as the sunshine sector in terms of its contribution to the growth and sustenance of the Indian agriculture. The COFI subcommittee on aquaculture (FAO-COFI, 2006) identified the following as major factors to create an enabling environment for aquaculture to develop enough to meet the demand for calories and protein through fish:

★ National and International market developments and access to markets,

★ Changing population and demography, seafood consumption habits and patterns, consumers preference and increasing consumers' purchasing power,

★ Technology development and improvement in management systems,

★ Improving public sector enabling environment, governance and institutions,

★ Access to services,

★ Adoption of environmental management practices for protection and sustainable use of aquatic resources,

★ Access to quality inputs in sufficient quantity,

★ Access to land and water resources,

★ Adequate physical infrastructure,

★ Ensuring food safety,

★ Skill development and capacity enhancement, and

★ Efficient communication and knowledge transfer.

The production of food is an over-riding priority, politically, economically and socially for the nation. Intensification and diversification in food production are two avenues to achieve the goal. The growth of aquaculture and its contribution to food supplies are dependent on its environmental interactions.

Thus, it is needless to say that the development of aquaculture has attained great importance so that the available natural resources may be exploited for the uplift of the common people. Considerable progress has been made in increasing per hectare yield by applying the modern/Hi-tech aquaculture methods at grass root level.

Chapter 2

Indian Fisheries Resources and their Classification

On the basis of exploitation and their distribution the fishery resources are broadly classified under two major groups such as:

I. Coastal fisheries or Inshore fisheries

II. Inland fisheries

Coastal Fisheries or Inshore Fisheries

The coastal fisheries contribute major share of total production of Indian's fish catches out of 6.4 million metric tones of India's total fish catches 3.4 million metric tones was estimated from the coastal fisheries during the year 2004-05.

India has vast coastal fisheries resources. The length of coastal line of India is estimated approximately 8129 km in length. The marine fish landing of the Indian coast mainly comprises of oil sardine followed by mackerel and prawns. The fishing season starts on set of monsoon and fishing is done by the mechanical vessels.

Inland Fisheries

When it is compared with the total expansion of ocean and sea, Inland water of globe seems very insignificant on the basis of ecological and geographical conditions. The inland waters are classified under two groups:

1. Lotic environment
2. Lentic environment

Lotic Environment

Under this running water series includes a forms of inland waters continuously in a definite direction namely streams rivers and estuaries. In the lotic environment a series of tiny rivulets gradually deep its head. Thus, in time extending its length and increasing its cross sections to the continuous process, ultimately produces a creck then finally a river. Hence lotic water generally stands as:

Streams ——→ Rivulets ——→ Brook ——→ Creck ——→ Rivers

In the lotic environment where continuously coming from up streams with fast current and high velocity.

Distribution of Lotic Environment

On the basis of physical and geographical conditions the lotic waters are grouped as follows:

(*a*) Streams

(*b*) Rivers

(*c*) Estuary

(*a*) Streams

It may be defined as a mass of water with its loads, moving more or less in definite pattern and how the course of least resistant towards the lower elevation, the streams are more and more apt to be highly turbid. The main source of water for the streams are heavy rain fall or melting of the winter's snow and major supply is from the ground water, primarily known as seepage on the basis of the continuity of the flow of the streams.

(*i*) Permanent Streams

The permanent streams receive their water mostly through seepages, springs and from subsurface of water. In the inside drainage area, the water level usually stands as a higher level than the flow of the streams.

(*ii*) Intermittent Streams

These streams receive their water primarily from surface run off. The run off is seasonal and the streams flow occurs during the wet period only.

(*iii*) Interrupted Streams

The stream which flow alternately on and below the surface is called interrupted streams. The surface is usually through coarse, sand or gravels. The flow of streams may be of three kinds:

1. Linear

2. Turbulent, and

3. Shooting.

(*b*) Rivers

India has a vast wealth of reverie resources. The rivers of India could be classified as follows:

(*i*) Himalayan rivers, and

(*ii*) Peninsular rivers

(i) Himalayan Rivers

Himalayan rivers include:

1. The Indus river system
2. The Gangetic river system, and
3. The Brahmputra river system.

(ii) Peninsular Rivers

Peninsular rivers include:

1. Narmada and Tapti river system
2. Mahanadi river system
3. Godawari, Krishna and Kaveri river system

The collective length of river, their main tributaries and canals run approximately 1,97,024 km.

(c) Estuary

An estuary has been defined as a body of water in which river water mixes with miserably diluted sea water. This is an ecotone or buffer zone between freshwater of rivers and salty sea's water. It has also been described as the wide mouth of the river or arm of the sea where the tide meets the river current or flow. The estuarine fisheries of the country is vast and have considerable scope for the development of both culture and capture fisheries. The total area of estuaries is 0.29 million ha.

In the Hoogly estuary, the tidal effect is felt upto 90 km from the mouth of the sea. In the Mahanadi estuary system, the tidal effect is up to 48 km. In general it is observed that marine fishes predominate on these systems.

Distribution of Lentic Environment

Lentic environment is known as standing water series or stagnant water series which includes all forms of inland waters such as lakes, ponds and swamps (Beels and Chours), in which the water motion is not that of continuous flow in a definite direction, and the movement of water occurs due to wave action, internal current on the vicinity of inlet and outlet. These are expressed as:

$$\text{Lakes} \longrightarrow \text{Ponds} \longrightarrow \text{Swamps}$$

1. Lakes

On the basis of lentic environment they are classified as:

(*i*) Natural lakes

(*ii*) Artificial lakes

The lakes defined as a body of standing water occupying a basin and lacking continuity with the sea or rivers. There are two types of lakes:

(*i*) Oligotrophic

(*ii*) Eutrophic

The first is unproductive lake having much depth where as second is productive lake having relatively less depth.

2. **Lagoons**

Lagoon may be defined as a large sheet of water surrounded and having a seasonal connection with the sea. These provide generally brakish water environment. In India about 0.29 million ha of brakish water area is available in the form of estuary, lakes and back waters. Of these Chilka lake, Pulicat lake and a Vambanad (Kerala) backwater form, all together 0.17 million ha. The main fisheries of lagoons are prawn, cat fishes, clupeids, mullets, prawns etc.

3. **Reservoirs**

In order to meet the deficit and to accelerate the economic growth of the country Central power and water Commission directed its attention to harness the potentialities of the great river system, giving large artificial lakes and reservoirs. These serves flood control, hydroelectric production, water supply to big cities, irrigation recreation and fish culture. According to the estimation, more than 295 reservoirs having the area of 3.15 million ha constituting a good source of fish production.

4. **Ponds and Tanks**

It may be defined as a shallow excavation of piece of land with possibility of holding water containing vegetations, which are used for fish culture. The total area of these ponds in the country is about 2.35 million ha.

5. **Brakishwater and Paddy Fields**

Among the coastal line of India there are vast extents of low lying areas which maintain more or less salinity throughout the year. They are unfit for agriculture purpose. The area of these brackish water, creeks and marshes are estimated about 1.24 million ha only out of which 0.6 million ha is utilized for fish culture. The practice of important impounded fisheries is followed in west Bengal in Sunderban area. About 10,000 ha of water area are utilized for fish culture which is known as Bhasa-badha.

Classification of Fisheries

On the basis of sources and geographical conditions, there are three types of fisheries known as:

1. Freshwater Fisheries: which includes cold water fisheries, and plain water fisheries.
2. Estuarine or Brakishwater fisheries.
3. Marine Fisheries

On the basis of exploitation and cultivation these fisheries are broadly classified into two groups:

(*a*) Capture fisheries

(*b*) Culture fisheries

(*a*) Capture Fisheries

The exploitation of naturally stocked fishes from natural waters is termed as capture fisheries namely, lakes, marine and estuary.

(*b*) Culture Fisheries

Culture fisheries basically constitute the collection of spawn from natural sources or from artificial sources and pond management is practised for nursing or rearing the seeds. Further these seeds are stocked in the ponds and reservoirs to obtain the marketable size fish by applying all technical measures as evolved to culture of fishes.

Culture fisheries is seasonal while capture fisheries is permanent. In culture fisheries human efforts are more predominant whereas in capture fisheries, nature plays a dominant role.

Chapter 3

Prospects and Perspectives of Fisheries

Fish culture in India is an age old practice. Temple tanks were used for stocking of fishes. In village areas tanks and ponds are mainly used for irrigations or for drinking water. By stocking fast growing carps, rich crop of fish can be harvested; otherwise the water body remains as fallow unproductive areas.

Fish is proteinous and easy to digest. In the village areas, where the people are suffering from malnutrition, fish culture solves the problem of malnutrition to a great extent. Generally 70 per cent people in India directly or indirectly depend on agriculture. Agriculture is a seasonal activity. Fish culture in village areas provides additional employment to the agriculture labours.

Fish production in the ponds was of the order of 600kg/ha/year. The reason for low production is non-stocking or less stocking of fast growing carp seed due to non-availability of quality seed in adequate quantity and unscientific management of ponds.

With the successful breading of major carps by hypophysation in 1957 by Hiralal Choudhary and K. H. Alikunhi and subsequent standardization of induced breeding technology, quality fish seed production has improved very much. Use of H.C.G., Homoeopathic drugs and Ovoprim etc. has resulted in furthering induced breeding technology to get assured results. Dry and wet bundh breeding of carps are also contributing to the success of fisheries.

Successful hatching of fish eggs is as much important as successful breeding of major carps. Uses of conventional hapas have not been giving satisfactory results. The reason is chocking of hapa cloth by silt. Crab cuts resulting in loss of spawn and rise of water temperature of ponds in which the hapas are fixed and subsequent mortality of span. Modern hatcheries revolutionsed the spawn production. Example are glass jar hatcheries, shirgur bin hatchery, CIFE- modern carp hatchery, Cement circular hatchery etc. successful hatching is one aspect, but rearing a good crop of fry is another aspect. By repeated experimentation, nursery management practices have been improvised and good crop of fry production is achieved.

The stocking of fingerlings per hectare is 8000 to 10000/ha and by stocking six species all the food niche of the ponds are properly utilized by the growing fish seed resulting in the production of about 10 to 12 tons of fish per hectare per annum by following scientific pond management practices.

By developing integrated fish culture technology and by recycling organic wastes, higher yield of fish is achieved with lesser financial expenditure. The entrepreneur is assured of getting not only good crop of fish but additional income from other systems of culture. In dairy-cum-fisheries, sale of milk provides additional income where as cattle dung is used as fertilizer for the pond.

In poultry-cum-fisheries, sales of eggs and chicken meats provide additional income whereas poultry litters can be used as a fertilizer for the pond. In piggery-cum-fisheries, pork provides additional income whereas pig dung is good manure for the pond.

In mixed farming system even the pond bunds are used for growing vegetables and fruits yielding plants which give additional income to the fish culturist. Napier grass is grown on the pond bunds which are used as green fodder to the cattle grass carp which results in better growth in the pond.

The patterns of phased manuring, artificial feeding and improved culture systems have increased fish production from the village ponds. There is growing awareness among villagers to adopt fish culture as a recognized occupation.

The fish yield from reservoirs in India on an average is 15 kg/ha/yr, which is very low. Before starting construction of reservoirs, no attention was paid to the improvement of fisheries. Hence, it resulted in difficulty in fishing and other management problems. Predators like catfishes and snake headed fishes of family *Channidae* etc. take a heavy morsel of young ones of carps. Further adequate stocking of required quantity of fingerlings is also needed. Removal of trash fish, eradication of aquatic weeds, observing closed season and allowing major carps to breed in the condition of reservoirs for auto stocking, clearing of submerged forest wood stumps for creating proper fishing grounds, proper screening of waste water etc. are some names of the improvement methods for increasing the yield of fish from reservoirs. Fish farms may be constructed in the near vicinity of reservoirs stocking. By following the above methods fish yield of reservoirs can be reasonably increased.

Indigenous methods of cage culture and running waste culture have to be standardized and popularized to utilize the river and canal system for culture. Adoption of hi-tech of aquaculture technique for very high yield is the order of tomorrow for Indian fisheries. Hi-tech aquaculture techniques need to be standardized, manpower trained and steps shall be taken to popularize the technology.

Chapter 4

Construction of Freshwater Fish Pond

The primary requirement in the construction of fish seed farms is proper formulation and planning of different types of ponds for rearing various stages of fish *i.e.* span, fry, fingerlings and table size fish or brood stock. The primary consideration in constructing the fish pond is to select a good site, for that the following characteristic should be kept in the mind:

1. The soil having water retentive quality
2. Assured supply of water throughout the year
3. Facility for self drainage of ponds.

For constructing a fish pond three factors are primarily taken into an account:

(*a*) Topography
(*b*) Soil type
(*c*) Water supply.

(*a*) Topography

Topography is the surface of the area. Ponds can be constructed on elevated area which has a gentle slope. It should be near the swamps or marshy lands or across narrow flowing streams or in an area adjoining them.

Best ponds can be formed on gentle terrain with self draining facility. In marshy area or swamps earth has to be piled up to form ponds of required size. Barrage ponds are made by building a series of dams in a narrow flowing stream or terraced site.

The area shall not be too elevated from the source of water supply as it involves pumping of water. It is desirable to get water supply by gravitational flow.

(b) Soil Type

Soil is very important factor. The soil shall not have vertical or lateral seepage. Good soils are heavy clay, silty clay, clay loam etc. rocky, sandy gravelly and lime stone areas.

A rough field method of testing good soil is to take a soil sample in hand and squeeze it in the farm of a ball. If it does not crumble after some handling and tilting it may be considered as satisfactory.

Under unavoidable conditions, when ponds are dug in loose soils, the pond bottom is treated with bentonits, clay or other soil sealants. In this case the soil is uniformly spread, then it is thoroughly disked with pond bed earth and compacted by a roller. A mixture or cut bits of straw, clay any cattle dung also can be spread on pond bed, disked with pad bed earth and compacted by a roller.

(c) Water Supply

Water supply may be from wells, tube wells, reservoirs, streams canals, spring, surface run off, artesian wells etc.

Because of elevated locations, reservoirs are perhaps best source of water supply. Springs' are also good sources, since they do not dry up during summer. 'Open type' *i.e.* spring 'spring ground' type where the water seeps into an area from the bottom which can be utilized for locating stock ponds and marketing ponds as they require to retain water for the entire year. Streams or canals are also satisfactory source of water if they meet the following points:

1. The flow is great enough to fill the ponds.
2. It maintains a fairly constant water level.
3. The streams are not subject to excessive flooding.
4. The water shed is well vegetated.
5. The stream carries slight silt load so that even during rainy periods the water remains clear.

When ponds are constructed near stream, generally they are constructed a little away so that the ponds are not over flooded and the water from the ponds does not seep into the stream.

Ponds filled with water by "Surface run off" are very common in upland areas. In the average rain fall, the ratio of pond and water shed area should be 1:10 ha. In forest areas where the rain water is retained considerably on the surface vegetation and leafy mat, the surface run off is reduced to about 50 per cent.

Jhingran (1991) gave certain requirements for the normal fish site under Indian conditions. They are:

Table 1

Sl.No.	Particulars	Requirements of a Normal Fish Farm Site
1.	Nature of terrain	Non rocky with at least two meter deep soil.
2.	Slope of terrain	Land should be level or gently sloping.
3.	Physical quality of the soil	Clay fraction should be about 90 per cent of the whole soil, stone and gravel not exceeding 10 per cent.
4.	Chemical quality of the soil	pH near neutral total nitrogen 0.1 per cent, total phosphorus, 0.1 per cent, organic carbon 1.0 per cent, Free carbon dioxide 5 per cent.
5.	Rate of rain fall in water	Should be less than 1 meter per annum.
6.	Water table	Should not be far below the pond bottom when the soil is not water retentive.
7.	Water supply	There should be a source of perennial water supply nearby to meet the requirement of the farm.
8.	Biological productivity	Average plankton production per m^3 should range between 10 ml and 20 ml.
9.	Cost of construction	The construction cost is not very high.

Construction of Pond

The fish culture operations commence with the construction of ponds. Specific types of ponds are required for the culture of particular species of fish and their life-history stages. Different types of ponds required for the culture of Indian major carps and other fishes are nursery, rearing and stocking ponds.

An area map of the site available is prepared for pond construction. A bench mark is made at a suitable place preferably on the highest point, then the levels at fixed points are noted, thus a contour map is prepared. This gives a precise idea of the earth to be excavated. Area to be demarked for store shed, Watchman quarters, approach roads, various types of ponds etc. can be decided, and accordingly a blue print showing all the items shall be prepared.

Rectangular ponds are preferable for easy management, L : B ratio may be kept at 1 : 4 to 1 : 6 depending on the size of pond. Management problems for netting of fish shall be kept in mind while deciding L : B ratio and also lead lift problem and sufficient run for suitable growth should be borne in mind.

Alikunhi (1957) has given the ratio between nurseries, rearing and stocking ponds are as follows:

Total area of a fish farm = 4.0 hectares

Nursery ponds = 0.2 hectares

Rearing ponds = 0.8 hectares

Stocking ponds = 3.0 hectares

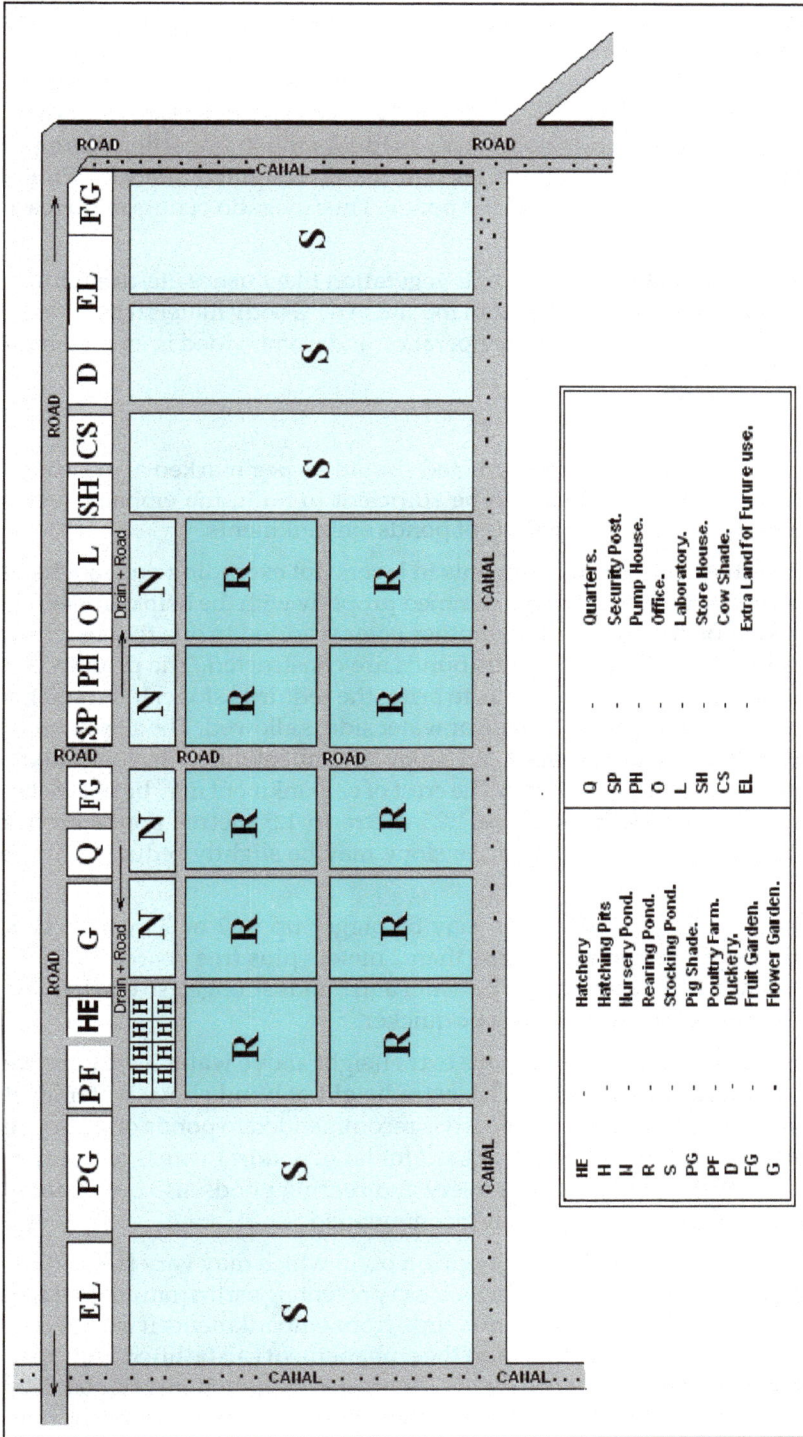

Plate 1: Proposed Model of an Ideal Fish Farm

HE	-	Hatchery
H	-	Hatching Pits
N	-	Nursery Pond.
R	-	Rearing Pond.
S	-	Stocking Pond.
PG	-	Pig Shade.
PF	-	Poultry Farm.
D	-	Duckery.
FG	-	Fruit Garden.
G	-	Flower Garden.

Q	-	Quarters.
SP	-	Security Post.
PH	-	Pump House.
O	-	Office.
L	-	Laboratory.
SH	-	Store House.
CS	-	Cow Shade.
EL	-	Extra Land for Furure use.

Under field conditions when a nursery is stocked at 5 million spawn per hectare and a 50 per cent survival up to fry stage is expected when the total fry yield will be 2.5 millions. At the stocking rate of 1.25 million fry per hectare in rearing ponds, the area required to rear 2.5 million fry is 2 hectares. With an expected 60 per cent survival up to fingerling stage the rearing ponds will yield 1.5 million fingerlings. Adopting the stocking rate of 5,000 fingerlings per hectare, the area required to stock 1.5 million fingerlings will be 300 hectares of stocking ponds. Thus the ratio of nursery to rearing to stock ponds should be 1:2:300.

Before any construction is started all vegetation like bushes etc. are removed. Roots and other foliage are removed from the site. Any woody material if exposed in the pond bottom would hamper netting operation and if embedded in embankments, may cause leaks.

Construction

The area where ponds to be constructed should be peg marked at appropriate locations. The earth excavated should be sufficient to build the embankments to minimize expenditure on constructions of ponds embankments.

The dry earth is laid on embankments in layers not exceeding one foot height. The clods are broken and the surface is rammed properly with the help of roller. Then water is sprinkled on the surface and another layer is added to one foot height and the process is repeated. Thus the ponds bunds are constructed. The profiles of the embankments are carefully drawn so as to bring the required slope. If the soil good quality clay a slop of 2:1 on the inner side or water side is allowed. The slop should be 3:1 for loamy, silty or sandy soils. A 2:1 slope of embankments is suggested for embankments of less than 0.6 m height. The crest of embankment may be of 1 meter. If the ponds are made in swampy areas 1.25 metre to 1.5 metres ponds crest are suggested. If ponds are stone packed the slope may be slightly reduced say 1.5:1 instead of 2:1.

The crest of table size fish ponds may be bigger up to 2 or 3 meters. This is required as the depth of pond is not less than 2 meters plus free board. Further for vehicle movement also the crest should be wider and bunds stronger, so that harvested fish and tackle and gear movement can be quicker.

All earthen embankments shall have extra height above water level to prevent waves and flood from overflowing. This extra height of bund above water level is known as "Free board". 60 cm of free board is recommended to ponds of water areas 0.44 to 1.2 hectare and about 90 cm free board for large ponds. In areas of heavy rain fall 60 cm free board is suggested to nursery and rearing ponds also but in areas of low average rain fall 30 cm free board is recommended.

Berm is generally left out before digging a pond which may vary from 30 cm to 90 cm depending on the size of pond. This helps in preventing earth from embankments slipping directly in to pond. When earth slips from embankments it settles on the berm which can be lifted with spades and the embankment can be lifted with spades and the embankment can be reshaped easily. The erosion of the bottom of embankment is also curtailed by providing suitable berm while constructing a pond. At the time of construction, the embankment should be at list 10 per cent higher than required, excluding the free board to provide for settlement of earth. The grasses should be applied on the bunds after the construction is over. The following water drainage system should be followed during the construction of the ponds.

(*i*) Drain Pipes

While constructing a pond a slope of 1 foot for every 100 linear feet provides for easy and quick drainage and the pipe should be large enough. For ponds up to 1.2 hectare pipe of 150 to 200 mm diameter should be suggested. The drain pipe shall be fixed at the lowest point of the pond for facilitating complete drainage. For bigger ponds, a spillway is also recommended. The area near the spillway shall have luxuriant growth grass as otherwise the earth will be eroded during flood situation resulting during peak rainy season. The materials used for pipes are asbestos, cement pipes, cast iron, fiber glass, Galvanized iron, concrete tiles etc. Suitable collars shall be provided at joins to prevent leakage.

(*ii*) Inlet Pipes

Inlet pipes shall be provided at least 150 cm above water level when the pond is full and shall be provided with screens to prevent entry of unwanted fish and escape of resident fish from pond.

In order to prevent erosion of embankment either grass or stones should be laid to serve as mat. Fencing may be provided to safeguard the farm area from stray cattle etc. Rats and crabs are a serious damage causing agents, Rats can be trapped and crabs may be killed by a suitable poison or by netting. Holes formed by the rats and crabs in the pond bunds shall be promptly repaired. To avoid the erosion the grass with clay shall be turfed at the marginal slope of newly formed embankments in order to check the erosion.

Clay particles from certain soil types make the water turbid thereby reducing productivity. In order to make the water clear agriculture gypsum at 192 kg per 1000 m^3 or say 200 kg per 0.1 ha of water volume may be applied. If not effective ¼ dose after 4-6 weeks may be tried. A second method is the used of super phosphate at 250 kg per hectare which gave satisfactory result. The use of super-phosphate gave the satisfactory result.

Reliable watchman shall be engaged to get the desired results from pond. If a watchman becomes a poacher the economics of fish culture will be adversely reflected. Trained dogs may be engaged for reliable watch and ward.

If freshwater ponds are well managed it is proven that one hectare water area gives more fish yield than one hectare of agriculture crop yield in economic consideration.

Chapter 5

Preparation and Management of Pond

The preparation of nursery, rearing and stocking ponds before releasing the stocking materials is an important step for successful rearing of carp spawn to fry, of carp fry to fingerlings and of carp fingerlings to table sized fishes. The processes for the preparation of all these ponds are more or less similar.

A. Preparation of Nursery Pond

For the preparation of a nursery pond following steps are followed:

I. Weed Clearance

Nursery ponds should be clear vegetation preferably by manual labour. This may be done during summer months, April and May. Pond should be kept free of weeds till the fry are harvested; a weed free pond facilitates free movement of the growing fry and is also conductive to production of their natural food.

II. Eradication of Predatory and Weed Fishes

One of the main causes of the mortality of spawn is the destruction caused by almost all types of small and large fishes present in the nursery pond. The commonly occuring predatory fishes in nursery pond are mostly *Channa puncatatus. C. marulius, C. straitus, C. gachua, Anabus testudineus, Wallago attu, Amphipinus chuchia etc.* predatory fishes are harmful to spawns as not only they complete with the latter for foods and space but also directly prey on them. Normally some of the predatory fishes breed in ponds a little earlier than the carps and their young ones feed vigorously on the plankton available in the ponds and grow so fast that when the carp spawn are released into nursery ponds, the predators are large enough to take a heavy toll of the spawn.

Weed fishes are the species of uneconomic, small sized fish that naturally occur or accidentally introduced in ponds along with carps spawn. The weed fishes breed in ponds in the summer months but are however, not as voracious as juveniles of predatory fishes. The complete removal of all such fish from the nursery pond is therefore, essential to ensure high survival and better growth of fry.

The customary methods of removing unwanted fishes from nursery ponds is by repeated drag netting, However, certain bottom dwelling fishes, *e.g.* Murrels, Climbing perch, Magur, Singhi, Cuchia etc. which burrow them in the mud are difficult to catch by repeated screenings. In such cases hooks and lines are used to catch large predators. The above stated methods, however remove the undesirable fishes at best partially dewatering of nurseries even by pumping, should be done to complete removal of unwanted fishes. If however, pumping is found uneconomical the nursery should be poisoned.

III. Poisoning

The objective of poisoning a nursery is merely to eradicate undesirable fish and the tender carp spawn can survive and grow. A suitable poison is one which is:

1. Effective in killing the target organism at fairly low doses.
2. Not injurious to the people and cattle who may use the water.
3. Not render the affected fish unsuitable for consumption.
4. Quickly nullified in water and leaves no cumulative adverse effect in pond.
5. Easily available and economical.

Several kinds of chemicals are used as fish poison which can be grouped as:

1. Plant derivatives
2. Chlorinated hydrocarbon, and
3. Organophosphates.

1.   Plant Derivatives

Derris powder with 5 per cent rate none content is most commonly used as fish poison. It acts as contact poison, damages the respiratory system and the fish dies. Other poisons are Tea seed cake, Mahua oil cake, Sugar cane jaggery etc.

2.   Chlorinated Hydrocarbon

The following hydrocarbons are mostly used for poisoning. They are:

(i)   *Aldrin*

It kills weeds and predatory fishes at 0.2 ppm.

(ii)   *Dieldrin*

It kills fishes at 0.01 ppm and prawns and insects at 0.5 ppm effectively.

(iii) Eldrin

Most poisonous and at 0.001 ppm concentration kills prawns and fishes effectively.

3.   Organophosphate

In India, Thiometon, DDVP and hosphomidon have been found successful for killing fishes on experimental basis. Mahua oil cake is an effective substance and acts as fish poison-cum-fertilizer. The oil cake may be applied at the rate of 2,500 kg/ha. or 250 ppm of pond water.

The required quantity may be mixed with water and scattered through out the pond. The fishes are affected in a few hrs. and can be netted out. The fishes killed by Mahua oil cake are fit for human consumption. The poisonous effect of the oil cake may persist for about 15 days. In the time, the minute organism *i.e.* plankton develop due to the manure property of the oil-cake.

IV. Pond Fertilization

The high level of fertility is considered essential in nursery ponds as to augment the production of zooplankton, the natural food of carp spawn. For this, the nursery ponds are first limited liming of the pond @200 to 300 kg/ha depending on the pH of the soil just after removal of weeds and unwanted fishes. Since, it does not only help towards better utilization of the fertilizers but also in disinfecting the nursery pond. Fertilization is done to encourage the growth of microscopic animals which form the natural food of the spawn. Pond is manured with fresh cow dung only, if mahua oil cake is used as the fish poison, 5000 kg of cow dung/ha. is needed other wise cow dung at 10000 kg/ha. has to be applied. Initial manuring of the nursery has to be done about 15 days before the anticipated date of stocking to ensure sustained production of plankton. For avoiding the risk of mortality of spawn due to depletion of O_2 in pond water, the cow dung is either used in the corners of the pond water or is kept in baskets under water. In India, cow dung is more favoured than inorganic (N-K-P) fertilizers, as the farmer leads to quicker production of large number of zooplanktons, the choice food organisms of Indian major Carps.

Fertilization of ponds by Mahua oil cake/cow dung may some times result in the development of microscopic plants and animals in week's time. When the growth of phytoplankton is too dense, the colour of the ponds becomes green/blue green. This is controlled by dissolving cow dung or some other organic matter or dyes in the surface water which prevails the penetration of light into deeper water and thus cause the bloom of die out. It is then removed by cloth net 1 or 2 days before stocking with spawn.

V.   Eradication of Aquatic Insects

Large number of predatory insect and its larvae which multiply profusely during the monsoon season, especially in manured ponds, can cause immense damage to many ponds as they fly from pond to pond. Thus, a pond which has been effectively cleaned of its insect population is soon repopulated with the insects. So, ponds should be cleared immediately before stocking. Their destructive role in carp nurseries has been described by many scientists. Most of the aquatic insects both in their larval or adult stages, in addition to preying directly upon carp spawn and fry and also complete with the latter for food.

Eleven orders of class insecta comprise the aquatic forms. Members of the order Coleoptera, Hemiptera and Odonata are important one. A great part of these insects can be controlled by slow dry netting with a fine meshed net about 3 or 4 days before stocking. To kill the remaining insects especially back swimmers (Notonectidas) by spraying an emulsion of cheap country soap solution with any expensive vegetable oil in the ratio of 56 oil: 18 soap kg/ha. is necessary. The emulsion first heated for some time and then sprayed over water surface. The spray should be done 12 to 24 hrs before stocking. A clean and dry day should be chosen for the spraying because in breezy and rainy day the spring does not give the satisfactory result.

VI. Stocking of Spawn

The container in which spawn is transported for stocking should be half submerged near the margin of the pond so as to gradually equalize the temperature of the water inside the container with that of pond. The container should be slowly dipped and filled in the pond, so that spawn are free to swim out of container in the nursery pond. It is better to release spawn in different sector of nursery. The operation should preferably be done during morning or late afternoon avoiding the mid day heat. The rate of stocking may be one to two million spawn per hectare. The appropriate time for stocking the nursery pond is when plankton biomass densely is sufficient upon the density of plankton the stocking rate varies from 12 to 20 lakhs/ha. However, much higher stocking densities 78,12,500 spawn per ha. as mentioned by Hora and Pillay (1962) are known to be adopted by fish farmers.

VII. Supplementary Feeding of Fry

The amount of natural fish food in the form of microscopic plants and animals present in the nursery pond has to be roughly assessed immediately before stocking is done. For this, 50 liters of water should be collected from various sports in pond and filtered through a plankton net made of fine muslin cloths. The filtrate is collected in a small tube of 2.54 cm (1 inch) diameter and a pinch of common salt is added to it. This will kill plankton, which will settle in the bottom. 1.0 to 2.0 ml of settled plankton of yellowish brown colour should be taken as satisfactory level of natural fish food in the nursery.

The spawn are voracious feeders and consequently the natural food would hardly last more than 4-5 days. To supplement the natural food, a feed prepared with finely powdered and sieved oil-cake/ground nut/mustered oil cake/coconut oil cake/seasem oil cake, fish meal are also to be applied from out side mixed with an equal weight of sieved rice or wheat bran may be used. The daily ratio of feeding nursery pond can be calculated on the basis of following schedule. 2800 gm/day for use of one million spawn.

1st to 5th day of nursery rearing: 2,800 gm/day/million spawn

6th to 10th day of nursery rearing: 4,200 gm/day/million spawn

11h to 15th day of nursery rearing: 5,600 gm/day/million spawn

An artificial feed comprising a mixture (called NPC mixture) of dried, finely powdered and sieved aquatic insects (back swimmers), small spans and shrimps and cheap pulses (cow pea) in the ratio of 5:3:3 according to Lakshman *et al.*, 1997; gives a better result in enhancing the survival and growth of spawn the mustered oil-cake rice bran mixture.

VIII. Harvesting of Fry

Under normal conditions fry raised as above, may be expected to grow to 25-30 mm in about 15 days with survivals ranging from 50-80 per cent. It is not advisable to keep the stock over crowded for a longtime. Since mortality due to infection and disease may occur. The fry should therefore be harvested within 15–20 days of stocking. Artificial breeding may be stopped a day before harvesting.

Two more crops of fry can be raised in a season in the same nursery pond after thoroughly cleaning the previous stock, in which case the pond may be fertilized with 2,000 kg/ha. of cow dung about a week before each subsequent stocking.

IX. Survival and Growth

The survival and growth of fry under the various management measures detailed above are generally quite satisfactory. Survival rates in nurseries have been registered as high as 88 per cent and an average of about 50 per cent with size of fry ranging between 20.0 and 30.0 mm. Normally the fry of individual species attain almost uniform growth through different species in the same pond may show different growth.

The growth rate and health condition of the individual species of the mixed lot of spawn are relatively better than those of pure seed stocked in separate pond.

B. Preparation of Rearing Ponds

The main target of managing a rearing pond is to culture maximum number of Indian major carp fry to fingerlings within shortest period of time. For this purpose, the rearing ponds are to be prepared more or less on the same lines of the nursery pond.

The rate of stocking with carp fry in suitable combinations and ratios and adequate supplementary feeding are essential for maximum survival and high production of fingerlings.

Experiment has been conducted at the pond culture division of CIFRI on rearing of carp fry to fingerlings stage in various combinations of Indian and exotic species. The results of the experiments on common carp conducted are given in Table 2.

C. Preparation of Stocking Ponds

These are large perennial water bodies varying from 0.2 to 2.00 ha in area, with an average depth of 1.5 to 2 meter.The general biological principles that govern the preparation of stocking pond are more or less similar to that of nursery ponds except a few exceptions. In large and deeper stocking ponds the poisoning may not be practicable for eradication of unwanted fishes and there is no known method to remove them completely.

Table 2

Sl.No.	Experiment Details	Pond I	Pond II	Pond III
1.	Rate of stocking/ha.	62,500	93,750	125,000
2.	Species combination and ratio	Catla 3 : Rohu 4 : Mrigal 1 : common carp 1	Catla 3 : Rohu 4 : Mrigal1 : common carp1	Catla 3 : Rohu 4 : Mrigal1 : common carp1
3.	Total no. of fingerlings harvested	53,650	78,800	95,900
4.	per cent survival	85.84 per cent	80.85 per cent	76.72 per cent
5.	Gross production/ ha/3 months	2,0550	2,511.75	3,486.12

In order to augment production of natural fish food organism in small stocking ponds, it is necessary to manure them in suitable installments with organic manure *i.e.* cow dung 20,000 to 25,000 kg/ha./yr and inorganic fertilizers, a mixture of ammonium sulphate, Single super phosphate and Calcium ammonium nitrate in the ratio of 11:5:1 at 1,000 to 1,500 kg/ha./yr.

Every pond can support a fish biomass only upto a certain weight limit. This is called the carrying capacity of the maximum standing crop. The carrying capacity of pond is intimating related to their natural productivity and management practices adopted, including fertilization, artificial breeding, the species stocked and their stocking rate.

It has been noticed that the over all production in mixed culture is 13-35 per cent higher than in monoculture. The number of fish to be stocked in a pond can be calculated by the following formula:

$$\text{No. of fish to be stocked per unit area} = \frac{\text{Total expected increase in weight}}{\text{Expected increase of weight of individual fish}} + \text{Mortality (not more than 10 per cent)}$$

Chapter 6
Pond Productivity

Lakes, reservoirs and ponds constitute a great source of inland fisheries in India. Lakes are naturally formed depression on the surface of the earth and get filled with water. A reservoir is an artificially created body of water by constructing a Bundh or a Dam across a river. The term pond denotes a very small shallow body of water. The productivity of a pond or a lake depends on the quality of water and soil. During production, radiant energy is transferred into chemical energy by photosynthetic and chemosynthtic producers. The rate of biomass accumulation is known as production and rate of production is known as productivity.

Types of Productivity

A. Primary Productivity

The rate at which energy is stared by green plant is called productivity. In other words it is called amount of energy stored per unit of time per area. It is further divided into:

(a) Gross Primary Productivity (GPP)

It can be defined as the total rate of photosynthesis including the organic matter used up in respiration during the measurement period. It is also known as "total photosynthesis" or "total assimilation".

(b) Net Primary Productivity (NPP)

In NPP the organic matter used up in respiration is excluded from the gross organic matter produced during photosynthesis. It is rate of storage of organic matter in plant tissues in excess of the respiratory utilization by the plants during the period of measurement. It is also called "apparent photosynthesis or net assimilation"

B. Secondary Productivity

The rates of energy storage at consumer level *i.e.,* heterotrophs are referred as secondary productivity. Since consumers only utilize food materials already produced appropriate respiratory loss and convert to different tissues by over all process. Secondary productivity should not be divided into GPP and NPP. There is still other type of productivity.

C. Net Productivity

It refers the rate of storage of heterotrophs or consumers in a unit period. Thus

Net Production = NPP – Consumption by heterotrophs

The productivity of a fish cultivation pond is understood to mean its capacity to produce cultivated fish. It includes maximum normal production which is highest possible production obtained and maintained under normal farming conditions.

Some ecosystem like ponds and lakes are essentially self sustaining. The fundamental steps in the operation of self contained ecosystem are:

(*i*) The reception of energy.

(*ii*) Production of organic material by producers.

(*iii*) Consumption of this material by consumers and its further elaboration.

(*iv*) Decomposition to inorganic compound and

(*v*) Its transformation into suitable forms for the nutrition of the producers.

Some environmental factors affecting production process in an ecosystem are:

(*a*) Solar radiation

(*b*) Mineral nutrition

(*c*) Biotic activity

(*d*) Impact of Human population

Although a natural condition of pH (6.5–7.5) is most favourable for productive pond, acid water and high alkaline conditions are unproductive. Total alkalinity below 200 ppm is indication of poor production and Do_2 above 6 ppm is suitable for a productive water body. Inorganic nitrogen above 0.2 ppm is favourable for productive ponds. These factors are very much important for the production of a pond.

Concept of Productivity

Three measures of productivity *i.e.* rate of production, rate of removal and standing crops are in common use and from the core of the concept of productivity in an ecosystem.

(*i*) Rate of Production

Rate of production relate to the amount of organic substance synthesized in a certain space per unit time. This is explained by primary productivity, secondary productivity and net productivity.

The rate of energy storage in the level of consumptions and decomposers are called rate of secondary production. Productivity is generally expressed in terms of gram or kilo calories per squire meter per day or year (PG = PN + R).

If gross production is equal to respiration (PG = R_1), no change in energy content results but when PG is less than R, biomass decreases and when PG is greater than R_1 there is an accumulation of biomass. The rate of primary production involves the cycle of growth, reproduction, death and decomposition of organism.

If the total solar energy enters a pond, a small portion of it is utilized during the process of photosynthesis by green plants and is stored in the form of carbohydrate chiefly in plant bodies. Not all the energy present in photosynthetic product takes part in the actual growth of plant. Since the process involved is not 100 per cent efficient, actual plant growth is represented by an amount of energy equal to the total carbohydrate formed minus (–) the respiration as is termed as the net primary production.

The secondary production constitutes the energy food actually assimilated by heterotrophic organism such as primary, secondary and tertiary consumers and decomposers.

The fishes depend mainly on phytoplankton and zooplanktons. Hence to have a better quality and quantity of fish, the primary productivity should be enhanced.

(*ii*) Rate of Removal

Various aspects of the biota removed and harvested from an ecosystem from the second measure of productivity *i.e.,* rate of removal. It refers to the yield or harvest from an ecosystem per unit of time. The emigration of animals or plant life, predation by terrestrial animals, outflow of organism, through effluent streams, loss of nutrients of sediments, commercial harvesting of fish are some of the ways of removal. Fish mortality in respect of production is only a portion of the total amount of material to keep itself removed from the area. However, if a community is to maintain itself, the loss of material must be compensated by adequate replenishment.

(*iii*) Standing Crop

The standing crop in the total biomass of the organism exist in the area at the time of observation and may be expressed as number of individuals as biomass of energy contained. The term standing crop is modified as "usable stock" in case of fish population. Usable stock is defined as the weight of all fish in a stock which is within the range of unusable size.

The measure of productivity does not include the time or rate, element involved in the formation of crop. The measurement of standing crop can be determined by only removal population density of an individual in the ecosystem at a given time frame. Thus a high standing crop has a high rate of production. The expression of members does not remove the time factor in production of crop. It may be long or short.

In the first place, in general, the natural production of the pond must be predetermined. It is indispensable to do this all from where natural production is the most important for this case. It will determine the total productivity. This is the case for both extensive and semi extensive forms but is less important for intensive fish production.

Scientific Methods of Estimation

Scientific method for determining natural productivity in fish cultivation in the ponds or in any other aquatic environment rely on the studies of aquatic flora and nutritive fauna. This method is well known but is less well known in its relationship to fish production.

There have been many studies devoted to the establishment of primary productivity or aquatic flora of different environment but works on secondary productivity or aquatic fauna and fish cultivated productivity are less humorous and impressive.

Empirical Methods of Evaluation

1. Leger-Huet Formula

The productivity of running water and of artificial ponds are given in following formula:

$$K = B \times L \times K \text{ (For running water)}$$

$$K = Na/10 \times B \times K \text{ (For artificial ponds)}$$

The annual productivity/km of a water course (K) is expressed in kilos in equal to the products of biogenic capacity (B) of the average width of the water course (L) and the productivity co-efficient (K)

$$Na = \text{Areas of stagnant water.}$$

Thus the different amounts of productivity of the natural productivity formula are the biogenic capacity, size and the co-efficient capacity.

(a) Biogenic Capacity

Biogenic capacity is the expression denoting the nutritive value of water examined for its feeding qualities for fish. It is designated by the abbreviation "B" and is expressed in a form called the scale of Biogenic capacity of which the degrees correspond individually to a given nutritive value from I to X. They are:

(i) Poor water–I to III

(ii) Average water–IV to VI

(iii) Rich Water–VII to X

(iv) Sterile water–O

The value and richness of pond present in a given median depends not only on this quality but also to their quality in relationship to the food required by the fish living in the food required by the fish living in the environment to be estimated. The biogenic capacity of a given aquatic environment depends in consequence directly on the quality of nutritive organisms which the fish seeks. The aquatic fauna depends on living or non living macroscopic and microscopic organisms, the latter being present as phytoplankton and also as biological cover.

It has been pointed out that certain physiographical characteristics influence the biogenic capacity. They can be grouped as physical, chemical and mechanical and they are themselves dependent on geographical and climatic characteristics.

(b) **Surface Area**

It goes without saying that the productivity of an aquatic median whether running water or still water is proportional to the extent and size of the surface of the water.

The surface area plays its part in the productivity formula of running water as well as for artificial ponds. In the later case, the surface area is expressed in "areas" (1 area = 100 m^2 and 1 hectare =100 areas) and is divided by 10. If the surface area is expressed in acres (1 acre = 100 m^2, 1 ha = 2.471 acres), the productivity is found by replacing the factor N/10 by 4-Nac *i.e.*, four times the no. of area.

For running water, the surface area factor also intervenes for width of the water course which is measured in meters and as the formula gives the productivity/km it then refers to 1000 meter in length.

2. Productivity Co-efficient

The productivity formula for certain artificial still water is derived from the formula of Leger.

The co-efficient "K" is composed of four secondary co-efficient designated as K_1, K_2, K_3 and K_4 which corresponds to respectively to temperature (K_1), chemical characteristics of water body (K_2), species of fish K_3 and age of fish (K_4). The product of these four secondary coefficients gives the value of K (productivity co-efficient).

$$K = K_1 \times K_2 \times K_3 \times K_4$$

K_1 = Average annual temperature

K_2 = Acidity and alkalinity of water

$10°C = K_1 = 1.0$ {acid water $K_2 = 1.00$}

$16°C = K_1 = 2.0$ {Alkaline water $K_2 = 1.5$}

$22°C = K_1 = 3.0$

$25°C = K_1 = 3.5$

K_3 = Species of fish

Cold water species $k_3 = 1.0$ {K_1 = Age of fish}

{Over 6 months, $K_1 = 1.0$}

Warm water species $k_3 = 2.0$ {under 6 months $K_1 = 1.5$}

The extreme value of productivity co-efficient K_w ($K = K_1 \times K_2 \times K_3 \times K_4$) under conditions mentioned are between 1.0–16.

Oxygen Measurement

There is definite equivalence between oxygen and food production. Oxygen production can be the basis for determining productivity of the water bodies.

Light and Dark Bottle Methods

This method is employed for measuring and estimating of O_2 production. Sample of water from different depths are placed in paired bottles and are covered with black tap or aluminium foil to avoid light penetration. The other water samples are fixed with reagents, so that the original oxygen concentration of each depth can be determined by Winkler's Volumetric Methods. Now the string of paired dark and light bottle is suspended in the pond, so that the samples are at the same depth from which they were drained.

At the end of 24 hours, the string of bottle is removed and the O_2 concentration in each sample is determined and compared with the concentration at the beginning. The decline of O_2 in the dark bottle indicates the amount of respiration by producers and consumers *i.e.* the total community in the water as O_2 changes in light bottle reflects the net result of O_2 consumed by respiration and O_2 produced by photosynthesis. Adding respiration and net production together or subtracting final O_2 concentration in the dark bottle from that in the light bottle gives an estimate of the total or gross food production for the 24 hours period. Then the oxygen is released in proportion to formation of the food.

The combination of light and dark bottle measures are the gross primary production and the light bottle measures are net community production.

There are two drawbacks of this method:

1. More recent researches have shown that there are many errors in the results.
2. This method obviously does not measure metabolism of the part of the community on the bottom.

From what has been said above, it shows a clear picture that to have a qualitative cum quantitative yield of fishes we must have the full idea of productivity. The high productivity means the increased dry weight and the increased energy content of fishes. However, we should search out some economical factors to induce the productivity which directly enhances the fish production. By knowing the productivity, we can adopt the method in the fish culture to have a very high yield.

Chapter 7
Biology of Indian Major Carps and Exotic Carps

A. Indian Major Carps

Labeo rohita, Catla catla and *Cirrhinus mrigala.*

1. *Labeo rohita*

Hindi Name: Rohu.

Distribution

Labeo rohita have a large distribution in freshwaters of India. Their high occurrence were reported in freshwater rivers, reservoirs and ponds of Punjab, Uttar Pradesh, Bihar, Madhya Pradesh, West Bengal, Assam, Orissa, Gujarat, Maharastra and Andhra Pradesh. It is also reported that they occur in freshwater in Pakistan, Bangladesh, and Burma.

Labeo rohita have also been exported from India to other countries. The details are as follows:

Table 3

Sl.No.	Country where Exported	Year of Export	Quantity (Fingerlings)	Sale/Gift
1.	Africa	1968	2000	Sale
2.	Japan	1961	6000	Gift
3.	Malaysia	1957	4,700	Sale
4.	Nepal	1957-58	95,000	Sale
5.	Philippines	1965	15,000	Sale
6.	USSR	1966	3000	Sale

Habit and Habitat

Rohu can thrive well in all freshwater below the altitude of 549 M.S.L. The minimum temperature 670 F is the lowest tolerance limit. Large fishes generally remain in deep waters, but for feeding, come to the shallow area. During monsoon season Rohu migrate for breeding in shallow areas.

Food and Feeding Habits

Rohu is a column feeder fish and starts feeding on natural food. In 5-7 day after hatching, it mainly feeds on vegetable debris, microscopic green planktons, decaying macro vegetables, detritus, and mud. Feeding intensity in adults is affected by the maturation of gonads and spewing of fish.

Sexual Dimorphism

Sexes can be easily differentiated. In Male, pectoral fin is slightly stouter and long (touches 8th–9th scale of the lateral line), while in female comparatively smaller (touches 6th–7th scale).

Ventral part of the pectoral fin is found rough in male, but in female, it remains smooth. During breeding season, on pressing the belly of male fish, oozes the white colour milt and in female ova comes out.

Maturity

Rohu attains the maturity towards the end of second year of the life in pond water. Gonads start maturation from the month of February and reaches on peak maturity during June and July. Male matures early than female.

Spawning

Rohu spawns once in a year but in southern area, same fish can breed twice in a year. Spawning season coincides with south west monsoon and north west monsoon. It breeds in rivers, reservoirs and bundh type tank where running water condition prevails. It depends on the temperature of water air, rains and cloudy weather. Thunder storms influence the natural breeding. Natural spawning generally occur in second and fourth phase of flood in the shallow marginal areas of rivers.

In confined water, fish easily breed by injecting pituitary gland extract.

Fertilization

Fertilization is totally external.

Fecundity

Fecundity of Rohu ranges from 2.5–3.0 lakhs/Kg body weight. Fecundity increases with the increase in length, weight and also the age of the fish.

Fertilized Eggs

Fertilized eggs of Rohu are oval, shining and demurral in nature. It is 3-4 mm in size, with a prominent fertilized nucleus.

Development

Development rate and incubation period of embryo depends on the temperature of water and dissolved oxygen. Low temperature increases the incubation period.

Optimum temperature ranges from 26°–28°C. At this temperature eggs hatch out in 14–16 hrs.

2. *Catla catla*

Hindi Name: Catla.

Distribution

As *Labeo rohita*.

Habit and Habitat

As *Labeo rohita, Catla catla* can tolerate slightly brackish water.

Food and Feeding Habits

Catla is a surface feeder fish and mainly feed on zooplanktons like crustaceans, rotifers and insects. Small quantities of green algae are also taken during the feeding.

Sexual Dimorphism:

Sexes are difficult to distinct externally except during breeding seasons when female shows bulged abdomen with radish colour. Its characters are the same as in *Labeo rohita*.

Maturity

The male matures in two years while the female in third year of its life.

Spawning

As in the case of *Labeo rohita*.

Fertilization

Fertilization of *Catla* ranges from 80000–1,20,000/Kg of body weight.

Fertilized Eggs

Spherical, but some are oval, transparent, light red in colour, non adhesion and floating. 5.0–6.0 mm in diameter.

Development

As in the case of *Labeo rohita*.

3. *Cirrhinus mrigala*

Distribution

As in the case of *Labeo rohita*.

Habit and Habitat

As in the case of *Labeo rohita* but it is a bottom dweller, large fish generally remains in deeper area.

Food and Feeding Habits

Due to bottom dwelling habit fish showed omnivorous habit. It browses mainly on decayed plants and animal matter, algae, detritus, mud etc.

Sexual Dimorphism
As in the case of *Labeo rohita*.

Maturity
As in the case of *Labeo rohita*.

Spawning
As in the case of *Labeo rohita*.

Fertilization
Fertilization: 1.4–1.5 lakhs/Kg body weight.

Fertilized Eggs
Spherical, transparent, non adhesive non floating and having a size 4.5–5.0 mm in diameter.

Development
As in the case of *Labeo rohita*.

B. Biology of Exotic Carps

1. Silver Carps (*Hypophthalmichthys molitrix*)

Distribution
A silver carp naturally occurs in the river system of south and central China and in the Amurbasin of Russia and associated countries.

The species has been introduced in many countries of the world including India. In India, the first consignment was brought from Japan in the year 1959, successfully cultured and its fingerlings were distributed to the other parts of the country.

Food and Feeding Habits
A silver carp is a surface feeder fish and mainly feed on unicellular algae. Fry and adult fish mainly subsist on Myxophyceae, Bacillariophyceae and Chlorophyceae.

Sexual Dimorphism
1. Male and Female can be identified easily.
2. Pectoral fin is rough in male white smooth in female.
3. In male milt oozes out just by pressing the belly while in female ova comes out.

Maturity
Silver carp attains maturity in two-three years. Males mature earlier than female.

Spawning
In India, full maturity is observed from the month of May and June. The breeding by hypophysation technique with stripping methods are generally used. H.C.G is found very successful for the breeding.

Fecundity
    The total numbers of eggs are eighty thousand to one lakh per kg of body weight has been obtained.

Fertilization
    Fertilization is totally external.

Fertilized Eggs
    Fertilized eggs of silver carps are pale bluish in colour, demarsal in nature, round and having a diameter of 4.2 to 4.8 mm.

Development
    Development rate and incubation period of embryo depends on the temperature of water, dissolved oxygen, low temperature, increased the incubation period. Optimum temperature ranged from 26–28°C. at this temperature eggs hatched out after 14-16 hours.

2.   Grass Carp (*Ctenopharyngodon idella*)

Distribution
    Grass carps occur in the river system of China and in the Amur basin of Russia. The species has been introduced in many countries of the world including India. In India the first consignment was brought from Hongkong in the year 1959, for the culture purposes specially to control the aquatic weeds.

Food and Feeding Habits
    Grass carps completely feed on aquatic macro-vegetations. Grass carp like the other major carps have a toothless mouth but has modified strong and specialized teeth for rasping and crushing the vegetation. Grass carp can consume 40-70 per cent of its own weight per day.

Sexual Dimorphism
    1. Male and Female can be identified easily.
    2. Pectoral fin is rough in male while smooth in female.
    3. In male milt oozes out just by pressing the belly while in female ova comes out.

Maturity
    In India, fish (male) attains maturity after 2^+ years, in female it is 3^+ years.

Spawning Season
    Spawning season of Grass carp is observed as in case of Silver carp. Fish easily breeds by the hypophysation technique using pituitary gland extract.

Fecundity
    The total numbers of eggs are 80000 to 100000/kg of body weight have been obtained.

Fertilization
Fertilization is totally external.

Fertilized Eggs
Fertilized eggs of Grass carps are pale yellowish prominent in colour, demersal in nature, round and having a diameter of 4.5 to 5.0 mm.

Development
Incubation period of embryos depends on the temperature it may ranges from 24–30 hrs. Optimum temperature requires from 26–28°C.

3. Common Carp (*Cyprinus carpio*)

Distribution
Common carp is a native of Asia specially China, but at present due to acclimatization of its variety, it occurs in Tropical and Temperate waters of the world.

Common carp has three varieties as follows:

1. *Cyprinus carpio* var. Communis–Scale carp. (Tropical water).
2. *Cyprinus carpio* var. Specularis–Mirror carp. (Temperate water).
3. *Cyprinus carpio* var. Pudus–Leather carp. (Temperate water).

The first consignment of common carp (specularis and nudus) was bought to India from Ceylon in 1937 and was stocked in Nilgiri hills. Another consignment of common carp (scale carp) was brought from Bangkok in 1957 and was stocked in Cuttack waters where it was successfully reared and bred. The fry and fingerlings were distributed to various state of India. At present common carp is available in almost all freshwater bodies in India.

Habit and Habitat
Common carp is a pond dwelling fish and is found at the bottom side.

Food and Feeding Habits
Common carp is a pond bottom dwelling fish and is found at the bottom of the pond. It is omnivorous in nature and feeds on bottom debris, sand and mud, nick of the aquatic weed like Hydrilla and Vallisanaria. This fish also found in the burrows in the embankments of the pond.

Sexual Dimorphism
1. As in the case of silver carp, Male and Female can be identified easily.
2. Pectoral fin is rough in male while smooth in female.
3. In male milt oozes out just pressing the belly while in female ova comes out.

Maturity
Common carp attains maturity in 6-8 months.

Breeding Season
Breeds 4-6 times in a year with a peak period in February to April and October to November.

Fecundity

It ranges from 1.5 to 2.1 lakh per Kg of body weight.

Fertilization

Fertilization is totally external.

Fertilized Eggs

Fertilized eggs of common carp are 1.0 to 2.0 mm in size, yellowish in colour, spherical in shape, adhesive in nature.

Development

Incubation period ranges from 48-72 hours, depending upon the water temperature. Newly hatched larva are attached with the leaves of the aquatic plant by means of the cement gland and remain in the same condition until the yolk is fully absorbed.

Chapter 8

Fish Seed Resources in India and Methods Applied for Seed Production

The diverse geographical and climatic conditions of India are reflected in the riverine resources of the country. The different river systems of country depending upon their individual ecological conditions, such as gradient, terrain, flow, depth, temperature, substrata etc. display variations in regard to the distribution and abundance of fish fauna. The rivers of the north originating from the Himalayan glaciers and snows are perennial and supports a rich commercial value of fishery in the plains of the three systems *viz.* Ganga, Brahmputra and Indus. The Ganga river system is the largest and contains the richest freshwater fish fauna in India. The East coast and West coast river systems differ from those of North and north-east not only that they flow through deep gorges but also in respect of their fauna and its fishery in the peninsular rivers is rather poor both in the upper and middle reaches.

In the Ganga river system the fish seed is collected in the form of eggs, spawn, fry and fingerlings. In case of Brahmputra river system only major carp spawn is available in the lower reaches and can be profitably exploited on a commercial basis compared to the Brahmputra river system, the Indus river system is rather rich. But only small portion is exploited. In the Narmada and Tapti river system there are some important spawn collection centres in Madhya Pradesh, shows the abundance of some species of carps. The Mahanadi river system has also a good potentiality in Orissa state.

Present Scenarios of Fish Spawning in River System

Construction of barriers across rivers and indiscriminate fishing and destruction of spawners during their breeding, migration and congregation at breeding ground is a common malpractice in the river system in India which may reduce the recruitment level. River valley projects have tended to change the riverine environmental efficiency of migration of fishes to breeding grounds and restricting the down flowing water thus affecting the flooding pattern in the rivers.

The use of rivers as the most convenient sewer system for industrial and human waste has resulted in the degradation of the riverine environment leading to distruction of breeding grounds in the rivers. The industrial and human wastes often result in to heavy pollution in the rivers. This is one of the most important factor affecting the production of spawn in these rivers. Destruction of juveniles of major carps lively to reduce spawner population level and ultimately the stock size. The future measures for conservation of stock, erratic monsoon and degree of precipitation affect the breeding opportunities and also affect the stock of other varieties of fishes that cause the reduction of the population balance towards unretarding levels in future.

Therefore, an urgent need of legislation and its implication to stop catching of fishes in pre and in monsoon period is required. Use of a definite size of net is also a must in the different river systems in India for the existence of the fishes in natural condition.

Fish Breeding in Bundhs

Bundhs are spell way type of perennial seasonal tanks or impoundment where riverine conditions are simulated during monsoon months. The Bundhs now is a well known source of carp seed contributor in the country. India needs a huge quantity of fish seed for stocking the above stated water resources. The increasing gap between demand and supply of fish seed can be narrowed to a large extent by adopting Bundh breeding techniques. Riverine spawn, still contributes, largely to seed production in India but due to its mixed quality, may find next place in market in coming years. Induced bred spawn is still meager as compared to demand. So the hopes for good quantity of quality seed largely rest on Bundh breeding. The West Bengal and Madhya Pradesh are the major source of producing fish seed from Bundhs.

A Bundhs is a local term meaning a Dyke–a Dyaked pond is also known as a Bundh. These are two types of Bundhs namely Wet Bundh and Dry Bundh. Generally the Wet Bundh is one which retains water, at listing a small portion through out the year. The other, Dry Bundh which is shallow seasonal pond retaining water for a few months. Hence, we should utilize these potential sources of quality fish seed to the maximum.

The known history of Bundh breeding is almost one century old, when Manu Teli a private pisciculturist successfully collected prawn from "Sorbati Bundh" Bankura district of West Bengal in 1882. However, the commercial collection of eggs from Bundhs was probably started in the years 1902 at Similapal belt of Bankura district. Systematic Dry Bundh breeding of Indian Major Carps came in existence in

Plate 2: Map of River System in India

1926 in the same district of West Bengal. At present there are as many as 1400 Bundhs in West Bengal 40 Bundhs in Madhya Pradesh and 135 Dry Bundhs in Rajasthan. However number of Bundhs in India may be much more if Bundhs of other states like Uttar Pradesh, Maharastra and Tamil Nadu etc. is considered.

Types of Bundhs

Depending on the nature and period of water retentively, the Bundhs are mainly grouped in two types–Dry Bundhs and Wet Bundhs. Dry Bundhs are seasonal tanks which dry up after rainy season and wet bundhs are the perennial tanks retaining water for many years. Both these types are discussed in detail:

A. Dry Bundhs

As stated above these are the seasonal ponds or tanks which are filled with rain water and dries up after rains. Any village tank or low laying area if embarked properly conserve as Dry Bundh for fish breeding. Technically speaking, any embarked low laying area having enough catchments area with proper gradient and outlet arrangements may be used as Dry Bundhs.

Site Location

Before locating the site of Dry Bundh we should consider major aspects like engineering and fish breeding. The layout of the land should be such where a good sized pond can be made with a small dam. Preferably any site with flat area surrounded on three sides by steep slopes and the fourth low laying side should be narrow as possible to reduce the cost of embankment may be selected for the Dry Bundh construction.

The engineering aspects include the area of Bundh proper, bottom contour, and water depth and embankment design.

Bundh Area

The area of Dry Bundh may be considerably. Conveniently and low laying pond with 0.5 ha.-1.0 ha. area can be converted into an ideal Dry Bundh. Bigger than this may pose managerial problems whereas smaller are insufficient and may prove uneconomical.

Bottom Counter

Bundh bottom contours should be taken into consideration. The bottom terrain should be gradually sloping towards the side in such a way that the lowest point lye at the bottom of Bundh, preferably somewhere near center so that a sluice can be provided there to drain out the whole water from Bundh whenever needed. Besides, there must be slight undulation in slop so that the marginal shallow dares can serve as breeding pockets.

Depth

Generally the Dry Bundhs are not very deep. Sometimes as less as 15-30 cm. depth of water inundation is sufficient for spawning of carps. However, the ideal depth of Dry Bundh may vary from 0.5-1 m. with shallow margin.

Embankment

Embankment should be made towards lowest side or from where water drains but of the pond. Ideal site for making embankment should be such that a small dam can serve the purpose to reduce the cost of embankment creation. Slopes of embankment should have 2:1 ratio on lower side of the dam and 2-3:1 towards the facing side of the pond. This may however vary a bit with the nature of soil or material used for making the dam. A minimum 4" width is desired at the top of embankment and a "free board" of 2" should be left for 1 ha. pond. Embankment height may vary from 4-8" depending upon pond condition.

There should be a guarded 'waste weir' and outlet fitted with fine meshed of 40 knot and nylon cloth or G.I. wire of 1-2 mm mesh. If possible, inlet (s) should also be guarded with such sieve to check the inflow of unwanted fishes and out flow of fish breeders, eggs, spawn etc.

Catchment's Area

Success of a Dry Bundh greatly depends on its catchments area. Scientists have given different options about the ratio between catchments area and Bundh area. It may varied from 5:1 to 25:1. However, an ideal ratio may be 10-5:1. Besides ratio of catchments area, its nature of vegetation and type of soil are also important factors contributing to success of a dry Bundh. Ideal catchments are a pasture land with slope gradient 1-4 per cent gritty soil or red laterite soil and moderate coverage of bushes, herbs and shrubs.

Bottom Soil

Bundh bottom soil must be gritty, alluvial, sandy, loam or clay in either order of preference. Sandy clay is generally not considered to be good from breeding viewpoint. Also it may pose problem while collecting the eggs from Bundh bottom. However, some scientists observed that soil type is not so much important.

Bottom Slop

The Bundh bottom slop should gradually deepen towards the center of the Bundh with mild undulating terrain to form shallow breeding grounds particularly towards margins.

Bottom Vegetation

It may indirectly influence breeding by stimulation breeders which rub their belly on marginal grass and also it checks the soil erosion. It is a matter of further study whether any particular type of vegetation bear any direct relation with fish breeding.

Rainfall

Rainfall plays a significant role on the gonadal development and thus in spawning of fishes. But it is generally agreed that for effective Bundh breeding sufficient run of water is required to influence the breeding intensity of carps and other fishes.

If the ratio between catchments and Bundh area is between 6-8:1 and the catchments cover is a pasture ground with red laterite soil and 2-2.5 per cent slope gradient, than even a minimum rainfall of 2 cm per day is sufficient for Bundh breeding.

Types of Dry Bundhs

On the basis of improvements in designs the Dry Bundhs can be divided into following three types:

1. *Primitive Type*

The traditional types of Dry Bundhs may be kept under this category. These are the simplest form of Dry Bundhs with minimum or no engineering background. Any low laying area with erection of small embankment and provision of inlet and outlet can be converted into a Dry Bundh, *e.g.* small seasonal village ponds, paddy fields and other such low laying areas which gets inundated with rains may be used as day Bundhs for breeding the fishes.

2. *Improved Types*

The most of the Dry Bundhs which are being constructed presently and which are working at present have improved designs considering various engineering aspects as stated above. Even embankments are made pacca or stone reverted with proper sluice gate, spill ways or waste-weirs, guarded inlets and outlets. Consideration is given to suitable catchments area, slop gradient, Bundh bottom contours and soil types etc.

In Bankura and Madinapure district of West Bangal the existing Bundhs are provided with a water reserves at high level to take water in Bundh during breeding time and thus efficiency of the Bundhs is increased. Breeder storage tanks conveniently of 3000 m² size are made to keep the nature breeders segregated sex-wise and ready before hand for fish breeding operations. Also in some Dry Bundhs a central pit of 30 × 10 × 1m. Depth is made to stock the breeders with first rain water which conveniently breed during next rains when water inundates the Bundh area.

These types of Bundhs are made either by fresh ponds or by improving the existing ponds as per convenience.

3. *Modern Type*

Very recently, some of the private entrepreneurs in West Bengal have constructed cemented Dry Bundhs, which they name as Bengal Bundhs. These are nothing but simple rectangular tank of size 75' × 25' × 1m. The first 50' bottom tank is at higher elevation with gentle slope (1 : 1) towards the rest 25' bottom which is at 1 m. lower level. So, almost a cistern of 25' × 25' × 1m. is made at posterior end of this tank. Thus when this tank is filled with water completely, the upper 50' portion will have a water depth of 1m and the lower 25' portion of 2m. There is a small 6-8" bump at the junction of these two portion and the demarking slop is not steep rather it is very gentle. Water is pumped from anterior and through pipes fitted with jets so that water gushes in the form of fine jets or showers towards the lower end of the Bundh. The breeders are kept in lower deeper portion with continuous water flow from upper to lower end, the breeders stimulated, and starts breeding.

However, detailed published information on constructional aspects and breeding operations of this type of Bundh is not available yet. But the private pisciculturist at Mogra (W.B.) has produced as much as 160-200 million spawn of Indian and exotic carp includes grass crap. This is a clear evidence of success and efficiency of Dry Bundh breeding.

Breeding Operations
In the traditional types of Bundhs at that time, no specific consideration was given to the breeder's ratio both sex wise and weight wise. Also no importance was given to keep the breeders segregated in advance before breeding. Only the good breeders were used to be released in Bundhs with onset of rains and the eggs were collected after breeding is over and hatched inside by earthen pits or by double hapas. Here both, the breeding and hatching results were usually very low. With improvement in breeding technology and Bundh designs, there was a break through in dry Bundh breeding potentials. The fishes are kept segregated in storage tanks near Bundh well in advance of breeding time. Then good mature breeders female and male in to 1:2 by number and 1:1 by weight are released @ 200 kg/ha in the Bundh which have received some rain water. Some of these breeders say 10 per cent are given fish pituitary injections to both male and female. After 4-6 hrs the fish starts breeding with out runoff water if the Bundh has already received sufficient rain water.

Problems of Dry Bundh Breeding
Sometimes the breeders are transported over long distances for Dry Bundh breeding. Such breeding suffers injuries during transport and fail to respond to the new environment and refuse to breed. The spawning ground should be weed free and complete drainage is also essential

B. Wet Bundh Breeding
Wet Bundh is a perennial pond, situated in the slope of a vast catchments area of undulating terrain, with proper embankment having an inlet towards the catchments area and an outlet at the opposite lower level. During summer, the deeper portion of the pond retains water containing carp breeders. After a heavy shower, freshwater from the catchments area rushes into the Bundh in the form of streamlets. The major portions of Bundh get submerged with water, the excess flow out through the outlet.

A modification of Midnapur type of Bundh is known as the Chittagong type where the mud walls are raised on the upland area for collection of rain water. The wall is cut at will to let the rain water collected towards the upland area to create conditions simulating riverine floods, suitable for the spawning of fish in the pond.

The Wet Bundh could be of any shape and size. Ordinarily these were understood to be small pond with a catchment area ranging 20-100 times but now even as a small body of water as a nursery pond (0.04 ha.) with no catchment area to as a large body of water as a reservoir of 300 ha have come so recognized Wet Bundhs. The shallow areas of Bundhs where the fish actually spawn are called moans. The outlet is protected by a bamboo fencing known as 'Chhera'. The flow of water through outlet can be controlled by blocking the chhera with straw and mud.

Spawning in Wet Bundh usually occurs after continuous heavy showers for days when large quantity of water rushes into the Bundh. At first, small size of fishes get stimulated to breed in the Bundh itself or in the area adjoining it. Bigger fishes spawn next in the same area. Though the fishes breed at any spot in the same Bundh, they may be beneficial to prepare spawning ground at different levels which could be flooded at night and during bright sun in the forenoon. The male chases the female. The 2 to 3 males to one female is the usual ratio of fishes. During the act of mating the female is held by the male, the latter bending its body round the female, rubbing, knocking and nudging her. At the climax of this activity the pair is seen to be locked in embrace their body twisted around each other with the fins erect and caudal fin quivering. The sex play lasts for a short time. The coiling and intertwining of two sexes erects pressure on the abdomen of the mating pair resulting in the extrusion of ova and exudation of milt. All eggs are not laid at one place and at one time, but at internal during which the pair keeps on moving. Fertilization in major carps is external. A large number of eggs accumulate in shallow water region giving pale whitish appearance. When spawning is over the spawning in the site, the spent fishes in Bundhs move to the deeper area in the Bundhs.

Seed and Hatching Collection from Dry Bundh or Wet Bundh

The eggs are collected by hapa or a piece of mosquito netting cloth after disturbing the water column. In West Bengal these are hatched in small and shallow earthen dug out pits with plastered mud walls with regular supply of water. Batteries of such pits are constructed by the side of the Bundh. In Madhya Pradesh hatching is done in double walled hatching hapa or cemented hatcheries. Now these days in some area the installation of portable hatchery unit near these Bundhs which is improve the hatching percentage and spawn recovering considerably.

Comparison Between Dry Bundh and Wet Bundh

Dry Bundh	Wet Bundh
1. Seasonal small pond.	1. Perennial large tank.
2. Controlled and can be easily manage.	2. Uncontrolled and difficult to managed.
3. Egg collection is easy.	3. Egg collection is difficult.
4. Desired quality spawn can be produced.	4. Mixed type of spawn is produced.
5. It can be operated 3-4 times in season.	5. Breeding is not controlled rather naturally.
6. More economical to Wet Bundh.	6. Less economical as compared to Dry Bundh.

Both Dry Bundh and Wet Bundh are good sources of fish seed production. If designed on the guidelines given above, the fish breeding is almost certain in Bundhs. No single factor is responsible for fish breeding in the Bundhs. Rather a combination of various factors like sufficient rainfall to create adequate currents by runoff water in Bundh, low temperature, fresh and well-oxygenated water, availability of suitable breeding grounds is responsible for successful breeding. However, if the fishes are given pituitary injection, the above stated favourable conditions may be passed.

Artificial water current may be created by pumping water in mini Dry Bundhs. Some more tips for successful fish breeding in Dry Bundhs are as follows:

1. Bottom should be covered with the sand or gravel, which may preferably be changed after each breeding operation or exposed to sun.
2. A combination of smaller and bigger breeders must be used as the smaller one gets stimulated easily and starts breeding first, following by bigger.
3. Water should be fresh and free from pollution and should not be contaminated with zooplankter. For mini Bundhs the ground water is an ideal source if used after proper aeration.

Chapter 9

Procurement of Brood Fishes and their Management

In fish seed farms, besides nurseries and rearing ponds, a section of bigger and deeper ponds are allotted for the culture of brood fish. The brood fishes may be grown from among the healthy fingerlings produced from the fry in the rearing ponds or they may be procured from the local tanks and transported to the fish seed farm a few months before the commencement of induced breeding season.

When the spawners are produced from the local ponds, the uninjured healthy stock is selected in the required proportion and is temporarily stored in the hapas fixed in the same pond till sufficient number of broad fishes are collected. While selecting the brood fish we have to see that:

1. Healthy and uninjured fishes are to be collected.
2. Breed fish should be collected in the proportion of 1:2 female to male ratio.
3. Fishes of almost equal size and weight range are to be collected.

When sufficient numbers of fishes are collected, they are to be transported by a suitable container to the fish farm. If the distance is short they can also be transported by cycles in a round tin carrier made of galvanized iron with thermo coal or sponges present in inner wall of container. Adult fish needs about 1 liter of water for 250 gm fish by weight for its safe transport.

Several cheap methods are engaged for the transport of adult fish. In Indonesia water tight tar coated plated bamboo basket of 150 litre capacity are used. In China 2 metre height × 2 metre diameter baskets and tubes are used for adult fish transport. In India splashes less tank of 1150 litres capacity, which has the facility for atmospheric air to be supplied while the fishes are being transported. And inside the splash less tank has a sponge quilting so that the fish is not injured by dashing against the metal containers during transport.

For the safe transport of breed fish, several sedatives are used. And the advantages of using sedatives in fish transport are:

1. Decreases rate of oxygen consumption and reduce rate of excretion of Carbondioxide, Ammonia and other toxic substances.
2. Controls excitability of fish thereby reducing the chance of injury.
3. Reducing the time required for handling.

Sedatives are used either by dissolving them in water medium and transporting fish in such media or by injecting the fish intramuscularly with certain narcotizing drugs.

The following anesthetic drugs are used for sedating the fish. The chemical is dissolved in water and the fishes are transported in that medium.

1. Sodium amytal 0.2 to 0.65 gm per gallon of water *i.e.* 52 to 172 ppm water is required for deep sedation of fish. But sodium amytal is antagonistic to calcium and hence not suitable for hard water. Only in soft water this chemical is used.
2. Tertiary amyl alcohol 2ml per gallon of water.
3. Chloral hydrate 3 to 3.5 gm per gallon of water. Narcotizing doses of the following chemicals are:
   (*a*) Urethane 100 ppm.
   (*b*) Thiouracil 100 ppm
   (*c*) Hydroxyquinadine 1.0 ppm.

The following chemicals are used for narcotizing the fish by intramuscular injection for transport:

1. Novacane 50 mg per 1 kg weight of the fish.
2. Barbital sodium 50 mg per 1 kg of the fish.
3. Amoberbital sodium 8 mg per 1 kg of the fish.

After transporting the fishes to the fish farm, before the fishes are stocked in the ponds, they are given a bath in chemicals to stop inducing of infectious diseases, predatory insects and aquatic weeds its seeds etc along with transported fish. A few chemicals are used for fish bath such as Acriflavin, Mthylene blue, Chloromycetine, Common salt, Copper sulphate and Potassium permanganate.

Generally in fish farms, brood fish ponds of 0.2 to 2 hectares size are used for culture of breeders which are easy to manage. Before they are filled with water, the ponds are desalted and the bunds are repaired whenever necessary. Lime at 200-300 kg shall also be applied depending upon the soil pH of the pond. The water shall be filled after screening so as not to allow any weed fish minnows to enter which will otherwise compete with the cultured fish for food, space and oxygen. Water shall be filled to a depth of 1.5 to 2 m for stocking brood fishes. The ponds are manured with suitable manures depending on the soil and water quality.

The following manures are recommended for fertilizing the tank in monthly or bimonthly installments. The requirements of manure are of cattle dung 20,000 to 25,000 kg/ha/annum. The quantity of manures to be applied may be decided depending on the soil and water quality. The raw cattle dung is applied in installment wise depending upon the requirement of the food.

The R.C.D is spread evenly on embankments before stocking and can be dropped directly into the water after stocking. The application of R.C.D is meant for enhancing the growth of natural food and helps in checking the seepages if any.

The inorganic fertilizer like N.P.K. (Di-Ammonium phosphate/super phosphate) if required can be used @100 kg/ha.

Fishes are to be stocked according to the carrying capacity of the stocking water. It is intimately related to the natural productivity of the pond and the management practices adopted including fertilization, artificial feeding etc. Generally 2000 to 2500 kg of adult fish per ha of water spread are recommended to be stocked. The species may include bottom, surface and column feeders so as not to leave any area unutilized. While stocking care has to be taken to see that almost equal size or weight group is introduced in the pond in the required male and female ratio. Undersized fish, yearlings or stock size fish shall not be introduced when the ponds are intended for breed fish management only. Breed fish need not be segregated except common carps.

Artificial feeding is to be restored with rice bran and ground nut oil cake at the rate of 1-2 per cent of body weight of fish. These things may be mixed in 1:1 or 2:1 ratio depending on the condition of the fish and gonadal development. Soyabean flakes or fish meal are also added in suitable dose in feed so that the protein level in the food is maintained from 30-40 per cent.

When algal bloom appear, feeding has to be stopped for a period. Early in the morning fish swim on the surface of water gasping for oxygen in distress when algal bloom persists. In such a case, water has to be let out and freshwater has to be introduced. The surface water has to be stirred with a stick etc to allow more atmospheric oxygen to dissolve which will be of immediate utility to the fish under distress. Steps have to be taken to control algal bloom by using algaecides like simazine, at 1 ppm or Copper Sulphate pellets at a dose of 30 kg/ha in six installments once in 3 days. Screening the water by using duck weed like lemna, axola and Wolfia may be tried. Sprinkling dung solution on water surface in the morning helps to cut down light penetration and thereby controlling the algal bloom. It is desirable to change the water once in 2-3 months. Addition of freshwater helps the fish to keep good health, when old water helps to remove some excreta, accumulated metabolites, obnoxious gases. Lethal substances and other repressive factor are washed out. During summer it is desirable to provide some shade if necessary by erecting pandals in the suitable corners attached with fruit producing big leaves trees for shade may be grown on embankments.

Due to constant fish movement, the bottom slush will often be disturbed. Some type of soil particles tend to remain in water column making pond turbid. Turbid water is unproductive, hence whenever turbidity are observed, steps have to be taken so that the water becomes clear by adding suitable chemicals. Agricultural gypsum at 2000 kg/ha, super phosphate at 200 kg/ha in suitable installments are recommended to control turbidity. Periodical trial meetings will ensure good health, good growth and condition of breeder's fish. Uninjured fish is necessary for getting successful result during the time of introducing breeding operation of fishes.

Chapter 10

Induced Breeding or Artificial Breeding

According to the nature of breeding habit and breeding requirements, the cultivated species of fishes may be divided into two main groups:

1. Pond breeding fish species that easily breed into ponds.
2. Species that do not easily breed in ponds and special methods are required to induce them to breed.

The fishes belonging to the second group do not ordinarily breed in ponds. These include the most important cultivated species like the Indian carps, Chinese carps, a few commercial cat fishes and the brackish water species like the mullet, milk fish etc.

Indian major carps are the most important food fishes which are extensively cultured in India. Although these species of fishes grow and attain maturity in ponds but they do not breed in this habitat. Consequently all requirements of carp seed have till recently been obtained mostly from riverine resources where these carps breed naturally. However, the seed obtained from nature is invariable having the mixture of economic and uneconomic fish seeds, the percentage of the later quite often being very high. There is also uncertainty regarding the unavailability of riverine fish seed, since natural spawning depends upon the factors, such as adequate and timely monsoon rains. Further, since the fish seed from the rivers are available at some specific centers located on rivers only, appreciable difficulty is encountered in transporting them to markets for sale. Thus, it is not always possible to have an assure supply of desirable species of seed at the required time from riverine resources. A problem of processing pure seed was successfully solved by culture division CIFRI at Cuttack in 1957-59 which lead to the development of special technique by which the pure carp seed can be produced from potential breeders by artificially inducing them to breed by injecting them with the extract of pituitary gland of the species of allied species. This technique of induced breeding is called "Hypophysation". This technique of hypophysation is being increasingly popular by fishries workers all over the country.

Pituitary Gland in Fishes

In fishes, the pituitary gland is formed in the form of a compact small mass of creamy-whitish in colour lying on the ventral side of the brain immediately behind the optic chiasma on the floor of the brain box. In shape, it may be spherical, cylindrical, oval or conical depending upon the species and the size and weight of the fish. For example a *Labeo rohita* in the weight range of 01 to 02 kg has a pituitary gland of 6.6 mg average weight, in 02 to 03 kg 10.3 mg, in 03 to 04 kg 15.2 mg, 04 to 05 kg 18.6 mg and like wise. The gland usually remains attached to the brain by a infandibular stalk which may be short, slender or moderately long. Sometimes, a distinct stalk may be totally missing. Gland having distinct stalk are classified as leptobasic type as in Cyprinidae type as in *Nandidae, Channidae* etc.

The fish gonadotrophins, in fact have not been studied so much in details as gonadotrophins in mammals. It is believed that like mammalian pituitary, fish pituitary also contains both FSH-like and LH-like gonadotrophins. These are however, different views regarding content and release of FSH and LH from fish pituitaries. In some fishes, two gonadotrophins have been described while in others only one type has been identified. Several attempts have been made in the recent past to isolate and purify the gonadotrophic hormones from the teleostean pituitary glands through chemical fractionation. Three different fractions could be obtained in the process of which the second fraction has been reported to be effective in inducing ovulation and spawning in fishes.

Collection of Pituitary Gland

Pituitary glands are collected from the ripe gravid fish of both sexes either belonging to the same species as the recipient or a closely related one. The most appropriate time or for collection of glands from the Indian major carp is just prior to or during breeding season. Since common carp (*Cyprinus carpio*) is a perennial breeder, its mature individuals can be obtained almost all round the year for the collection of glands. The glands are usually preferred to be collected from ice-preserved specimens.

Several techniques are adopted from the collection of pituitary glands in different countries. In India the commonly adopted technique of gland collection is by chopping off the scalp of the fish skull by an oblique stroke of a butcher's knife. After the scalp is removed the grey matter and fatty substances lying over the brain are gently cleaned with a piece of cotton. The brain thus exposed is carefully lifted out by detaching it from the nerves. In the majority of the Cyprinids, when brain is lifted, the gland is left behind on the floor of the brain box. The membrane Dura meter covers the gland. It is removed by a fine needle and forceps. The exposed gland is then picked up intact without causing any damage because damaged and broken glands result in loss of potency.

Glands are also collected through foramen magnum. It is, in fact, a much easier method of gland removal which is commonly practised by the professionals for mass scale collection in crowded and noisy fish markets. In this method of gland collection, the fish is required. The skull of fish cut with a sharp butcher's knife or a hand saw. The brain after being exposed is lifted and the pituitary gland is removed. In Indian fish market where cut fishes are often available, the glands are taken out from behind through the foramen magnum after changing the brain tissue.

Preservation of Pituitary Gland

The glands after collection are immediately put in absolute alcohol for defecating and dehydration. After 24 hrs, the glands are washed with absolute alcohol and kept again in fresh bottles and stored either at room temperature or in a refrigerator.

Acetone is also a good preservative. In this method, soon after collection, glands are kept in fresh acetone or in dry ice-chilled acetone inside a refrigerator at 10°C for 36 to 48 hrs. During this period, the acetone is changed two to three times at about 8 to 12 hrs intervals for proper defecating and dehydration. The glands are then kept out of acetone, put on a Filter paper and allowed to dry at room temperature for one hrs. They are then stored in a refrigerator at 10°C. Preferably in a desiccator charged with calcium chloride or any other drying agents. The preservation of glands in acetone is largely practised in Russian countries and the United State of America.

Preparation of Gland Extract

At the time of injection, the required quantity of glands are taken out and kept on a filter paper for alcohol to evaporate. The glands are then macerated in a tissue homogenizer by adding a measured quantity of distilled water. The concentration of the extract is usually kept in the range of 1 mg to 4 mg of gland per 0.1 ml of distilled water. After homogenization, the suspension is pouring, the homogenizer should be thoroughly shaken so that the settled particles of the gland at the bottom of the tube get mixed up with the solution and come into the centrifuge tube. The extract in the tube is next centrifuged and the supernatant fluid is drawn into a hypodermic syringe for injection.

The pituitary extract can also be prepared in bulk and preserved in glycerin in the ratio of 1 part of extract : 2 parts of glycerin before the fish breeding season so that botheration of preparing extract every time before injection is avoided. The stock extract should always be stored in a refrigerator or in ice.

Determination of Dose for Injection

Determination of right dose of gland extract depends mainly on the proper stage of sexual maturity of the breeders. In usual practice, female alone is injected with a stimulating dose of 2 to 3 mg/kg weight followed by a second dose of 5 to 8 mg/kg weight after an interval of 6 hrs, the males are given a low dose of 2 to 3 mg/kg body weight at the time of second injection to female.

Slight alterations in dose may occur depending on the stage of maturity of the breeders as well as environmental factors. When the males are not ripe enough, they also are injected with a stimulating dose of 1 mg/kg as in the case of the female (Alikunhi *et al.*, 1964).

Breeding Environment

The breeding hapa is a rectangular box shaped container made of mosquito net cloth, and are closed on all sides excepting one through which spawns are introduced or taken out. The breeding hapa is fixed with bamboo poles in a pond, river or any water shed.

Selection and Maintenance of Brooders

The brooders are selected well in advance and kept in breeding ponds. At the commencement of the rainy season, male and female are segregated in separate pond. Proper identification of sexes and stage of maturity are important requisites for good results. A fully ripe male oozes milt at slight pressure on the abdomen, while a fully ripe female has a soft rounded abdomen and swollen ventral side. Healthy breeders weighing 1.5 to 5.0 kg are preferably chosen for breeding. To keep the breeders in healthy condition they are often fed with mustard oil cake and rice bran for a few months prior to breeding season.

Spawning

After about 2 hrs of the second injection the breeders start swimming actively excited and restless and try to jump out. Spawning usually occurs within 3-4 hrs after the second injection. Slight splashing of water is observed just before spawning due to chasing of the female by males. In case the fishes do not spawn, then a second injection with a slightly higher dose is given. Spawning is ordinarily expected in another 3 to 6 hrs.

The major carps normally breed once in a year, either naturally or through induced breeding during monsoon. It may spawn twice within an interval of about two months. Almost equal quantities of eggs are obtained at each of the two spawning, thereby doubling the production of spawn.

Stripping

The females are considered to be fit for stripping only when the abdomen becomes very soft and stream of eggs oozes out freely on gentle pressure on the abdomen. The eggs are immediately stripped on to a clear, moist enamel tray and are mixed with the milt stripped from the males. A little water is added in the tray and the milt thoroughly mixed with the eggs by continuously tilting the tray. After some times, the fertilized eggs are thoroughly washed with pond water and excess of milt and dirt are washed out. The eggs are then transferred to large enamel trays with sufficient water for swelling and hatching hapa.

Hatching of Eggs

Eggs are hatched inside a series of hatching hapas tide to bamboo poles in the marginal water of pond. The hatching hapa consists of two parts, one filled inside

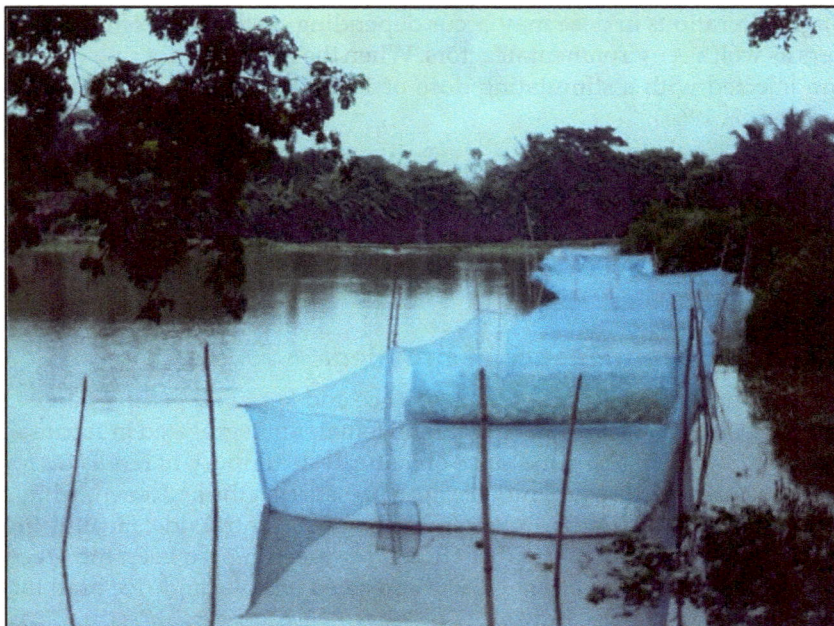

Plate 3
Figure 1: Hatching Hapas

the other is made up of thick meshed cloth, whereas the inner one is made up of round meshed mosquito net cloths. Eggs are stored uniformly all over the inner hatching hapa. The majority of eggs hatch out within 14-18 hrs after fertilization. After hatching, hatchlings escape to the outer hatching hapa through the hole of inner hapa. The hatchlings are left in outer hapa till the third day after hatching. Then the span is collected and stocked in nursery pond. Hatching depends to certain extent on the hydrological and climatic condition. Temperature range of 24°C–37°C has been found to be suitable for the spawning. The cool, windy and rainy days give better results.

External Factors Concerned with Induced Breeding of Fishes

The environmental factors such as light, temperature, pH, water condition, weather etc. are known to play important role in stimulating the release of pituitary gonadotrophin within the organism and thereby controlling the reproduction of fishes.

Early maturation and spawning of fishes as a result of enhanced photoperiodic regimes has been reported. Delayed maturation and spawning of some species of fishes are reported from the northern latitudes. Again, these are optimum temperature ranges for induced breeding of activated fishes and critical temperature limits above and below which fish will not reproduce.

Other factors responsible for spawning of carps in India are rain, flood, sudden increase in water level, suitable substratum, increase in dissolved O_2 content and pH of water are also considered important for spawning of major carps.

Use of Natural and Synthetic Hormones in Induced Breeding of Fishes

Induced breeding of fishes with the pituitary injection is becoming very popular and provides a reliable method of obtaining pure seed for pisciculture. Investigators have realized the necessity of finding other dependable agent to replace the pituitary extract for induced breeding. A few purified natural and synthetic hormones have been tested successfully.

Mammalian hormones have been used for ovulation and spawning in fishes and have good results for maturation of gonads in fishes. A mixture of chorionic gonadotrophin and mammalian pituitary extract known as synatorin has also been found in combination with fish pituitary.

The crude or pituitary purified human pregnancy urine as well as highly purified proportions of HCG has been used successfully in several fishes. Beside the above certain gonadal steroid are also known to be good ovulating agents.

Problem of Hypophysation

The following problem are encountered in the hypophysation of fishes:

1. The foremost need is isolation and determination of active ingredient of pituitary hormones. It is responsible for breeding of different species of fish. Such a development may lead to the synthesis of the fish gonadotrophic hormone.

2. There is need to acquire through understanding of the physiological conditions of natural fish pituitary gonadotrophin in relation to the season, age and sex development of donor species.

3. Suitable medium for preserving and storing the pituitary extracts for future use as these would help in salting up pituitary bank.

4. It is necessary to understood cytogenetically fish maturity and ovulation brought about by the administration of pituitary hormone.

5. Search for effective high pituitary substitutes both natural and synthetic should be vigorously continued so as to reduce the increasing demand for pituitary gland.

6. Little is known about the effects of different types of natural food and artificial feed on the growth maturation and fecundity of various culturable species of fishes. The role of food on the development of pituitary itself needs through investigation.

7. Selective breeding for stock improvement of a given species of culturable fish and hybridization for incorporating in the progeny, desirable parental or new qualities of high conservation value, fast growth, nutritive value palatiblity, maturation, fecundity etc. are rich dividends from

hypophysation. While hypophysation by trial and error may lead to practical results in field it would be necessary to understand cytogenetically the process of hybridization so as to evolve pure breed hybrids and have predictable results.

Factor Influencing Induced Breeding:

Favourable cool weather *i.e.* temperature in the range of 24°C to 31°C, cloudy and rainy period are good. Flowing water is preferred. Turbidity should be within 100 to 1000 mg^{-1}. Light drizzling following heavy rains is ideal.

Failures are mostly due to incorrect choice of breeders, wrong doses of pituitary extract and unfavourable climatic conditions. Hot, sultry or sunny days are not suitable for undertaking induced breeding.

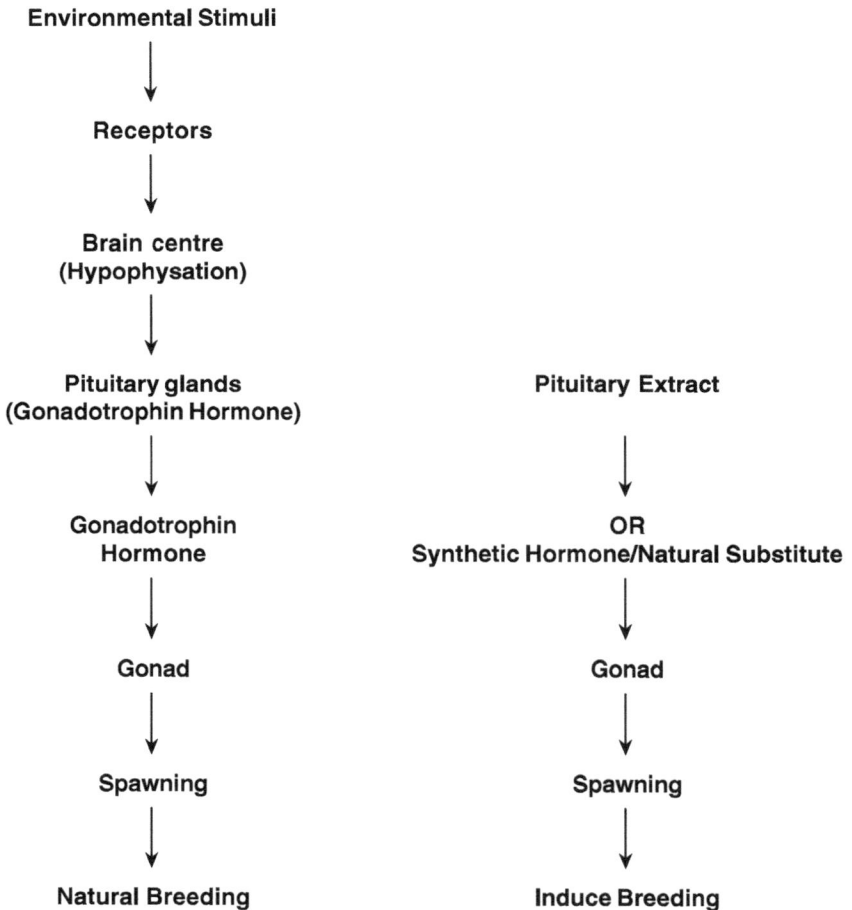

```
        Environmental Stimuli
                 │
                 ↓
             Receptors
                 │
                 ↓
           Brain centre
          (Hypophysation)
                 │
                 ↓
        Pituitary glands              Pituitary Extract
      (Gonadotrophin Hormone)
                 │                            │
                 ↓                            ↓
         Gonadotrophin                       OR
            Hormone          Synthetic Hormone/Natural Substitute
                 │                            │
                 ↓                            ↓
              Gonad                         Gonad
                 │                            │
                 ↓                            ↓
            Spawning                      Spawning
                 │                            │
                 ↓                            ↓
        Natural Breeding              Induce Breeding
```

Events in Natural Breeding and in Induce Breeding

Chapter 11
Induced Breeding of Indian and Exotic Carps

History of induced fish breeding goes back to the year of 1930, when Houssay for the first time reported the use of fish pituitary glands for fish breeding which was used in Brazil in the year 1937. In India the first unsuccessful attempt was made by Khan (1938) using mammalian pituitary extracts. And later on the first successful attempt was reported by Chaudhri and Alikunhi in the year 1957 on Indian Major Carps. It was reported to be effective in exotic Chinese carps also. Since then, the technique of induced fish breeding has been perfected to a great extent.

Indian major carps generally do not breed in confined water. They breed annually once during the monsoon in the upper reaches of rivers and tail ends of certain reservoirs where riverine conditions prevail. Also they are found to breed naturally in Bundhs breeding is gaining momentum in Madhya Pradesh and West Bengal. Scarcity of pure variety of fish seed in ample quantities has forced the scientists to find a way to breed the fish artificially by injecting with fish pituitary glands.

Method of Induced Breeding

Collection and Preservation of Pituitary Gland

Pituitary glands are taken out from a freshly killed fish. Glands can also be collected from a fish market from severed fish heads. Glands of common carps are considered to be good over major carps or Chinese carps. Pituitary gland is located in the ventral region of the brain in a cavity called sella turcica and covered by a membrane called duramater. Glands are preferred to be preserved in absolute alcohol because preservation is reported to be less durable.

Selection of Brood Fish and Formation of Breeding Set

For selecting a male, the fish is held gently with ventral side facing upward and the abdomen is pressed slightly near the ventral region, free oozing of milt with the slight pressure indicates mature condition of the male fish. Female with soft and bulging abdomen and slightly swollen with pinkish ventral are considered to be good. Usually a breeding set consists of two males and one female. But the ratio of weight is to be 1:1. In mass breeding practices, less number of males has also proved successful. Before injecting the fish, they are kept inside the breeding pool or in hapha to acclimatize to the changed environment.

Preparation of Dosage

Glands of a known weight is taken out from absolute alcohol and dried over a filter paper. And gland is put inside a tissue homogenizer and macerated with little distilled water. The Homogenized gland is further diluted to make the desired volume. The dilution is done at such a rate that each 0.1 ml contains 2-4 mg extract. The solution is configured using a hand operated centrifuge. The supernatant liquid is used for injection. Fish pituitary extract can be preserved in glycerine for long time use.

Determination of Doses

The problem involved in standardization of doses is of very complex nature, because the dose of gonadotrophin would depend upon a variety of factors including the size and stage of sexual maturity of the recipient fish, potency of glands at the time of its collection, freshness, stage of maturity of the donor fish. Land Preservation of the glands and also the ecological factor of the breeding environment at the time of taking up the induced fish breeding operation. However, following doses were reported successful in Indian major carps and also in Chinese carps.

Indian Major Carps

The ripe female fish of *Catla catla, Labeo rohita, and Cirrhinus mrigala* are injected twice at an interval of 4-6 hrs. The first dose is of 2-3 mg/kg body weight, and the second dose is of 4-8 mg/kg. Males are administered 2-3 mg/kg of body weight at the time of second injection to the female fish.

Common Carps, Grass Carps and Silver Carps

In these types of fishes slightly higher doses of pituitary extract is required. The first dose is of 3-4 mg/kg and second dose is of 7-10 mg/kg of body weight. The males are administered 3-6 mg/kg of body weight at the time of second injection to female fish.

(a) Method of Injection

(i) *Intraperitoneal Injection*

Injection is given in the peritoneal cavity of the fish. Specially in carps, there are soft regions at the base of both the pectoral and pelvic fins, through which the injection needle is inserted quite easily. In this method there is no danger to damage the internal organs and the injected fish may even die out of shock.

(ii) Intramuscular Injections

Intramuscular injection is ordinarily administered in the region of the caudal peduncle or below the dorsal fins but above the lateral line. While injecting, the needle is first injected under a scale parallel to body of the fish and then the muscle is pierced through quickly at an angle of 45°. A 2 ml hypodermic syringe with 0.5 ml graduations, preferably with a locking arrangement is convenient to use. The thickness and length of the needle depends on the size of the recipient fish. Generally number 22 needle is used for fish weighing 1-3 Kg and number 19 for larger ones. For smaller fishes below 1 Kg, needle number 24 is considered adequate.

(iii) Natural Spawning and Stripping

In natural spawning, eggs are released by the female and are fertilized by the male automatically. But in stripping method, the female milt is added afterwards to fertilize the eggs. Natural spawning is very common in Indian major carps but in Chinese carps, stripping method is used particularly in silver carps. HCG had been reported to result a natural spawning in silver carps. Usually the Indian Major carps spawn within 4-8 hours after the final injection to females and males. Chinese carps are ready for striping in about 6-8 hours and much earlier under suitable ecological conditions. In stripping methods, the eggs are stripped inside enamel tray and later on sufficient milt is poured over it. The eggs and the milt are thoroughly mixed with a feather and by rotating the contents by hand to ensure proper fertilization of the eggs. After thorough mixing, for about 5 minutes, some water is poured into the tray and the mixing of the content is continued. The excess of milt, blood clots and other extraneous matter is completely removed through washing stripping methods and they are of two types:

1. *Dry Method*: Fish is gently wiped and eggs are stripped in a dry enamel tray over which the milt is poured and followed as above.

2. *Wet Method*: Physiological salt solution (0.3 per cent) is used to receive the milt first and later the eggs are stripped. Water can also be used but the entire process of mixing has to be completed very fast.

Factors Responsible for Successful Spawning for Carp Breeding

In natural environment according to Khan (1945), Choudhary (1963) and Alikunhi *et al.* (1964), Indian major carps breed within a temperature range of 24-31°C. Hora (1945), Saha *et al.* (1957) and Jhingran (1968) are of the view that fresh rain water and flooded condition in a tank are the primary factors in triggering the spawning the carps. Successful spawning in majority of fishes has been induced in cloudy and rainy days, specially after heavy showers. Fish breeding has been possible under the controlled condition of hatchery even without any rainfall. Some of the scientists found that the raise of temperature gained by water while traversing across heated soil has got a paramount importance, and without the raise in temperature to a certain degree, no spawning is possible. Some fisheries scientists observed that aroma produced when water comes in contact of heated soil stimulates the fish spawning as fishes possess a remarkable sense of odour perception.

Sundaraja and Goswami (1974) has divided the annual reproductive cycle of teleosts specifically *H. fossilis* in following four periods:

I.   Preparatory Period   –   (Feb–April)
II. Pre-spawning period   –   (May–June)
III. Spawning Period   –   (July–August)
IV. Post Spawning Period   –   (Sept–January)

Gonadotropins play very important role during pre-spawning and spawning period. Pre-spawning period is associated with the increasing photoperiod and spawning phase is associated with environmental factors prevailing during the monsoon season. Spawning phase includes maturation and ovulation. In induced breeding operation. Injecting gonadotropin as fish pituitary gland is to help in maturation and ovulation.

However, an ideal environment and factor required during induced breeding operations are:

1. Selection of healthy fully matured ripe breeders in prime condition.
2. Collection of pituitary glands from fully gravid fishes.
3. Proper assessment of dosage of pituitary hormone.
4. A preferable optimum water temperature between 24–31°C.
5. Cool, rainy days are more conductive to fish breeding (Hapa beeding).
6. Dissolve O_2 level of 6-8 ppm.

Recent Advances in Induced Fish Breeding

Since the technique of induced fish breeding has become very popular throughout the country, so the availability of pituitary gland has become a problem. Efforts are being made to develop an alternate of fish pituitary gland which is effective, cheap, easily available and easy to adopt by fish farmers.

(a) Human Chronic Gonadotropin (HGC)

It is a protein hormone, produced by the placenta and extracted through the urine during pregnancy. It is more or less similar in character and functions to LH and FSH. It has been demonstrated by fishery Scientists that crude HCG of active about 301 i.w./mg is effective in silver carp breeding it is about 20 mg/kg of female and 5mg/kg of male. In Indian major carps a combination of 60-80 per cent HCG and 40-20 per cent PG results in successful breeding.

(b) Natrium Muriaticum

It is a homoeopathic drug used to stimulate quick ovulation in women. *Natrium muriaticum* @ 0.5ml/female weighing 1-2 kg resulted in oozing condition of the female fish in 12 hours of injection. All the female breeds successfully with a very low knockout dose of 2-4 mg of Pituitary extract per kg body weight.

Efforts are being made to synthesize pure hormones to standardize induced breeding techniques in fishes and to make the method further east and adoptable at a low cost.

(c) **Analogue of LH-RH-A**

LH-RH-A is also used for artificial breeding.

(d) **Ovaprim-C**

Superactive analogue of LH-RH, called Gonadotropin releasing harmone. It is 17 times more effective. For IMC 0.3–0.5ml/Kg is the most effective dose.

Common Carp Breeding

For successful fish farming, required good qualities of fish seed of desired species are required in adequate quantity. The ideal qualities of fish which can be grown in ponds, tanks and reservoirs are:

1. Easily available
2. Fast growing, and
3. Diseases resistances etc. for stocking.

In our country, major carps are extensively cultured in Inland waters. But these fish breed in only rivers in the rainy season and so the seed problems are great. But the Bangkok strain common carp breeds in small ponds and possesses all good characteristics for culturing in ponds and tanks.

Age and Size at Maturity and Fecundity

Under normal pond condition the fish becomes sexually matured when it is about is 6 to 8 months old. In same board, males mature earlier than the females by two months before. Minimum size at first maturity may be 15-20 cm with 80-170 gm weight. The 1 kg matured female fish gives 1.3 to 2.3 lakhs of eggs.

Breeding Season

The fish breeds throughout the year but having peak period from January-March and July–August month. Male sexually mature throughout the year. The common carp having capacity to breed several times in a year with successive intervals of 2-3 months.

Breeding

Ripe female starts with bulging of belly which is soft to touch. Fully matured fish is having pinkish red ventral surface. Males are comparatively thinner than females and ooze out white milt.

It is necessary to segregate the male and female before the season. The segregated good and healthy males and females are regularly fed on artificially with a mixture of oil cake and rice barn (1:1) @ 2-3 per cent of their body weight which may also use some proteinous diet with it.

Under natural as well as artificial condition, breeding ordinarily takes place at night or in the morning time. In artificial breeding, the fishes are kept in hapa or cisterns. A spawning female is usually followed by several males, wild splashing and frequent jumping are characteristics of the spawning activity. The pattern of matting is similar to that of major carps. Although the actual process of mating is difficult to observe as because of wild splashing. Ripen male can be used repeatedly.

The hapas are rectangular cloth and the hapas size is of 2 × 1 × 1 metre or 3 × 1.5 × 1 metre or 4 × 2 × 1 metre. The eggs of common carp is adhesive in nature and attached with weeds like Hydrilla and Najar and they are also kept into the spawning pool/Haper as an egg collectors. Somewhere coconut fibers are also occasionally used. With one fully matured female fish is kept with two or three males. The weight should be more than a male which are kept. The ration being egg collectors and fish 1 kg fish: 2.5 kg weed.

Hatching

After spawning the hatching is carried out in hatching hapas which are similar to the major carps. If fertilization is over 95 per cent than weeds will nearly go up to 1,00000 eggs. Now weeds should be shifted into hatching hapas sized 2 × 1 × 1 m. The incubation period may vary from 36 to 72 hrs depending upon on the temperature of the water as well. The eggs are then kept for three days. After absorbing the Yolk sac completely, the spawn start swimming activity which is a sign that they are ready for stocking in nursery ponds.

Chapter 12

Development and Management of Carp Hatchery System

In earlier days, availability of fish seed was entirely dependent on wild catch from rivers and large reservoirs during rainy seasons. This was accomplished by several problems. Apart from being the mixed type; seed collection itself is a troublesome task. Large number of predator and uneconomical varieties are used to be mixed along with the cultivated species of fish. Overall fish breeding is entirely depended on rains. No rain means no seed. To overcome all these uncertainties and collection problems, several types of fish hatcheries are there namely glass jar hatchery, Galvanized iron hatchery, shirgur hatchery, Chinese hatchery and modern carp hatchery etc.

Modern carp hatchery is the latest system for successful breeding and hatching developed by Central Institute of Fisheries Education, Mumbai. Hatchery has been tested in various agro-climatic conditions of different states of India including Madhya Pradesh, Gujarat, Haryana, Rajesthan, Andhra Pradesh, Utter Pradesh, Tamil Nadu, Maharastra and Union Territory of Delhi, and has been successful. Major advantage of this system is that it can be worked out in climatic conditions and is possible to install in any part in any short duration. Successful breeding and hatching results have been obtained in the controlled system of modern carp hatchery even without no rain in semi-conditions of Haryana. A series of modified system has been designed and had been named as CIFE D-80, D-81 and so on.

The system is divided into three units:

1. First unit,
2. Second unit, and
3. Third unit.

First Unit

It includes a tube well and cements cisterns/Filters. Ground water which is fresh and cool is pumped into cement cistern through a diesel or electric pump. Silt particles get settled at the bottom and oxygen level of water also increases gradually due to splashing of water.

Second Unit

It includes an overhead tank and a cooling tower. Freshwater from the cement cistern on filter is pumped into the overhead tank with the help of a pump. Oxygen level of water increases as water falls from a constant height. Sometime water is passed through a cooling tower. When ground water temperature is too high and is not considered suitable for successful hatchery operation.

Third Unit

Third unit embraces following three components:

(*a*) Breeding Pool

(*b*) Hatchery Jars

(*c*) Spawning

(*a*)   Breeding Pool

Size of the breeding pool depends on the weight of the fish. It varies from 6′ × 3′ size for smaller fish and 12′ × 3′ for larger fish. Rectangular/Circular cement cistern may also be used for the same purpose. It receives good quality freshwater from over head tanks. Oxygen contents of the water increases from 1.8 mg/l to 8.0 mg/l in breeding pools. Temperature is also maintained between 25–28°C. In this system, a moderate current of water is also maintained. Matured male and female brooders are brought in the pool and are injected with Pituitary gland extracts.

(*b*)   Hatchery Jars

It is the main component of the hatchery, where hatching takes place. It is of two parts, one outer another inner. Outer jar is made up of LDPE (Low density polythene). Inner is a mosquito netting jacket fitted in an aluminium frame. Basic shape of the inner is almost similar to the outer jacket. Inner jacket is put inside the outer. LPDE jar and fertilized eggs are loaded in the inner jacket. Mosquito netting facilities smooth flow of hatched out young ones from hatchery jar. It receives water from the overhead tank through an inlet at lower side which comes out from an outlet near the surface. Regular flow of water is maintained at an optimum rate so to provide continuous removal of metabolic wastes.

D-80 type hatchery jars are of flat bottom. D-81 type hatchery jars are of conical bottom. It provides better circulation to eggs. D-85 type jars are of conical bottom, with a holding capacity of 0.8-1.0 million eggs. Water flow is maintained at 10-12 l/m.

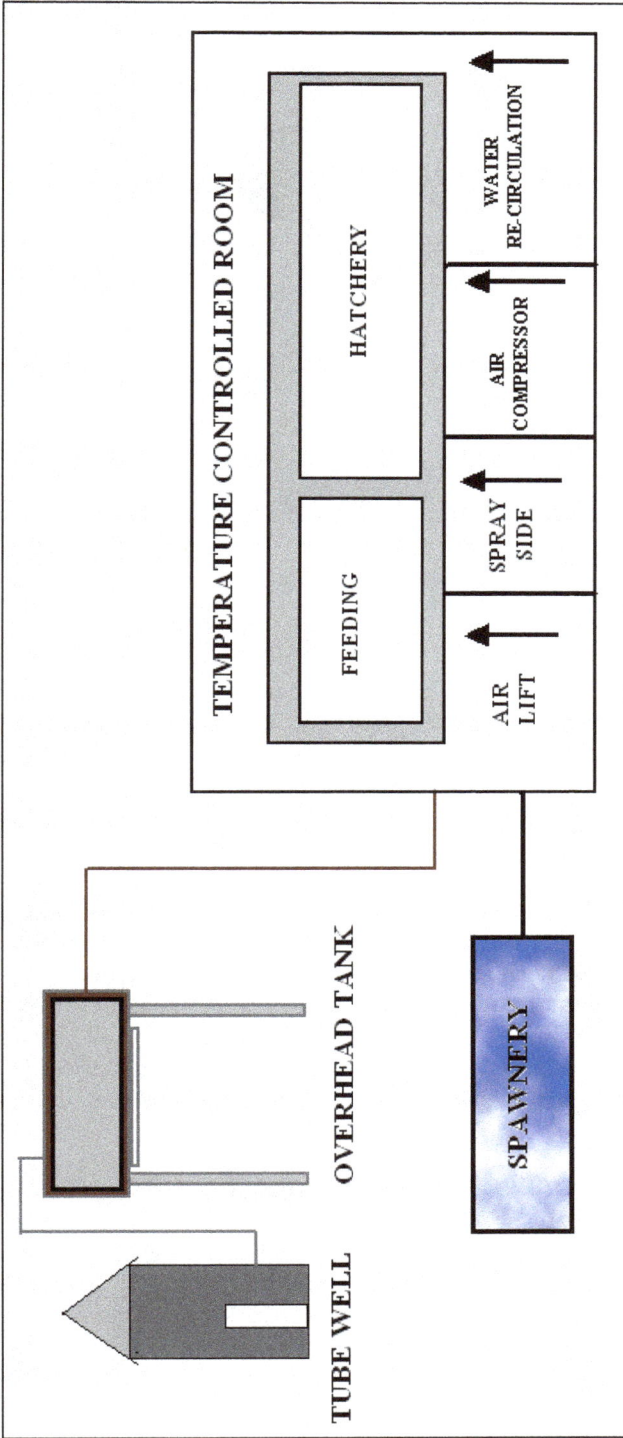

Plate 4: Model of D-81 Hatchery

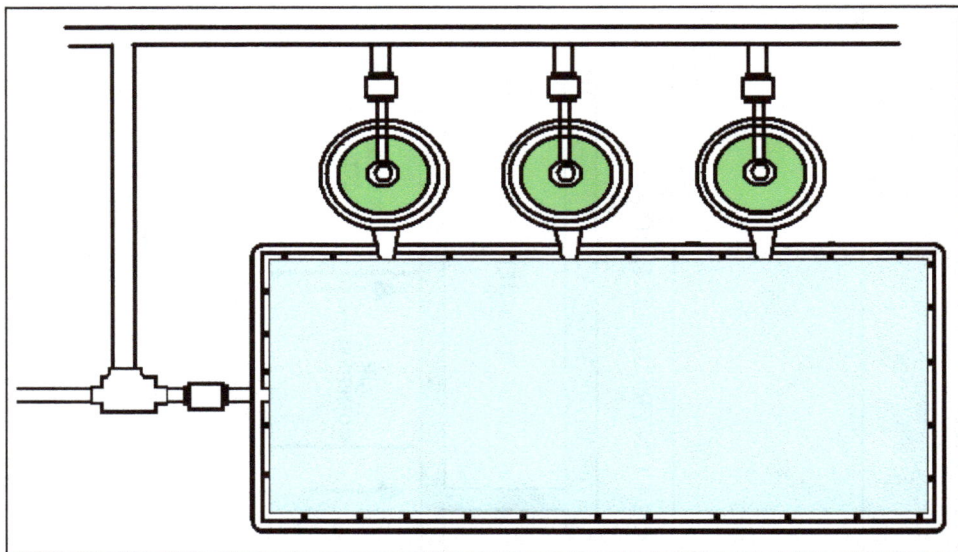

Plate 5: LDPE Vertical Jar Hatchery (D-81 Model)

(*c*)   **Spawning**

Same types of pools as used for breeding are used as spawnery. Hatchings are collected in this tank and are kept for 72 hrs. Water is circulated at a low speed; water spray and aeration are also provided to ensure cooling and high oxygenation to maintained optimum conditions for growth. A 20 mm PVC flexible or rigid pipe is arranged around the periphery of the pool, with small holes in it, which provides water spray.

Model CIFE D-86

It is designed to attend mass hatching at reservoir site. It is a pool of 8" × 3" size or 12" × 3" size, provided with reversal number of egg sacs. Egg sac is a mosquito netting jacket fitted on an aluminium frame with 2 to 3 floats. It ensures balanced floating of the egg sacs. Water circulation is maintained with air lift technique. Compressed air is passed through the system to support oxygenation. Freshwater supply is maintained regularly to allow continuous removal of metabolites.

Air Lifts Circulation

It requires a 2.5' long PVC pipe of 20 mm diameter and a band of the same size. Dimensions of the pipe may vary depending on the size of water body. Bend is fitted on the long pipe and is hanged in the water in such a manner that opening of the bend is half dipped in water. Compressed air enters in the long pipe just near the lower end. It lifts the water from bottom layers, which comes out from upper and opening. System provides uplifting of bottom layered water which is deficient in oxygen to the upper surface and mild circulation of water is also maintained.

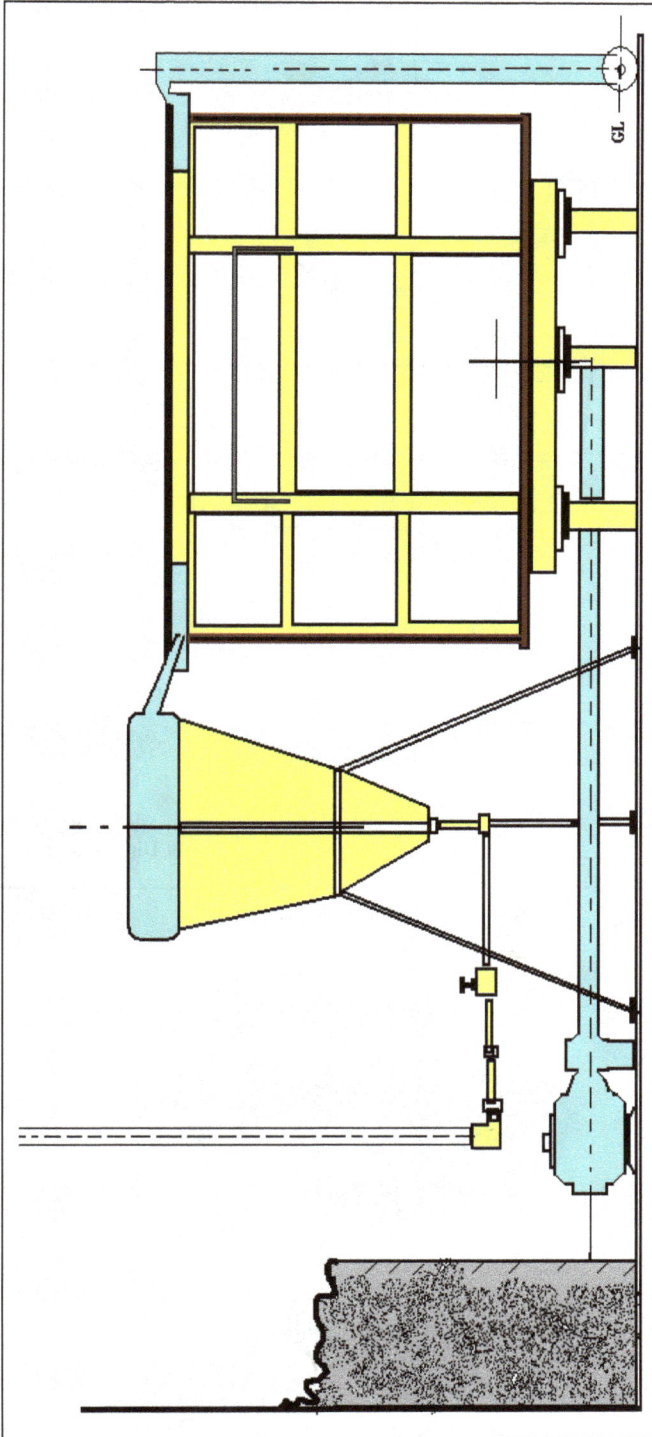

Plate 6: LDPE Vertical Jar Hatchery Side Elevation

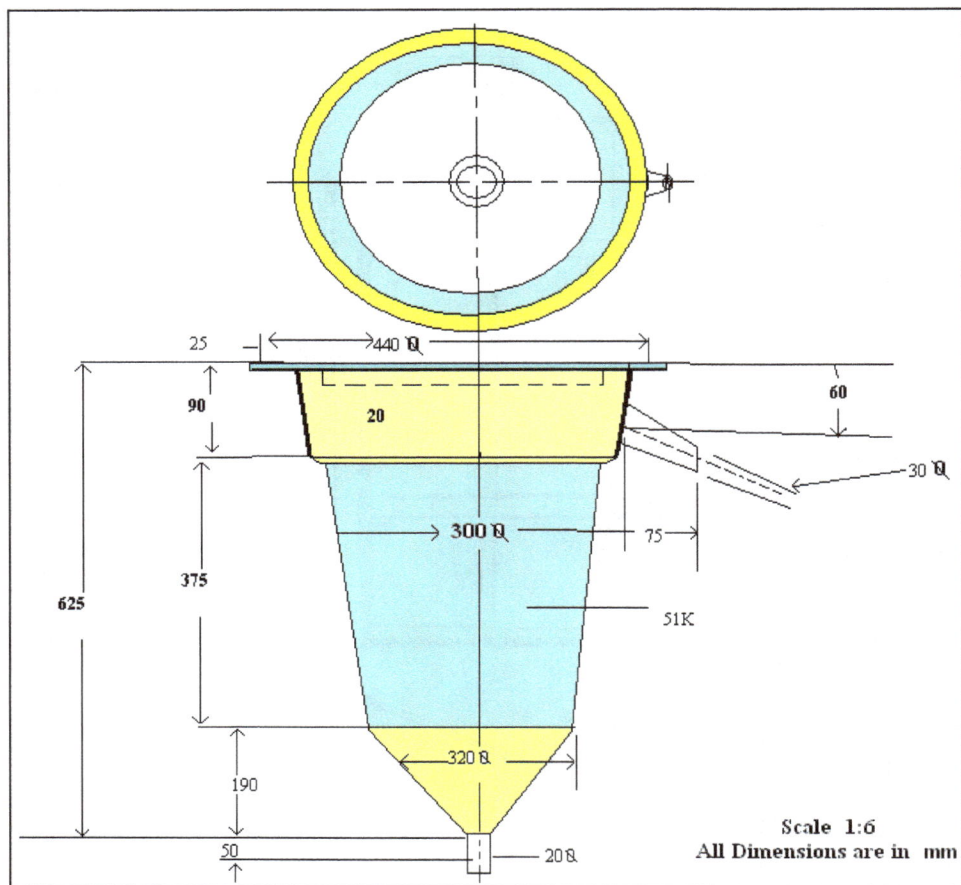

Plate 7: Vertical Jar Hatchery

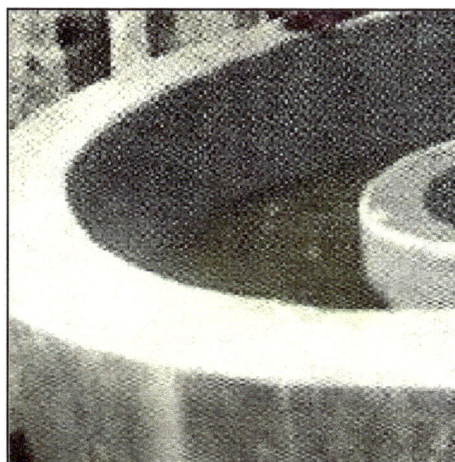

Figure: Circular Hatchery

Advantage of Modern Carp Hatchery

1. Easy to install and operate.
2. Transportable.
3. Low cost.
4. Easy to maintain controlled environmental conditions.
5. High hatching success.
6. Breeding is successful even without rains in any agro climatic conditions.
7. Long lasting, since made of LDPE.
8. Men power requirement is low.

Chapter 13

Oxygen Uptake in Early Developmental Stages of Indian Major Carps

Metabolism has been called the fire of life (Kleiber, 1961) and metabolic rate is the measure of the fuel consumed and heat evolved by that fire. Metabolic rate is increased by a number of different processes such as physical activity, food processing and tissue synthesis and it is also dependent upon the developmental stages and body weight of the animal. Utilization of oxygen is considered to be a valid measure of metabolic rate in animals. Moss and Scott (1961) and Beamish (1964) suggested that metabolic rate in relation to body weight in fishes are specific for different species. Slope values ranging for 0.45 to 1.0 have been reported to show the relationship between oxygen uptake and body weight by various workers.

A great deal of work on the relationship between oxygen uptake and body weight in the Indian air-breathing fishes have been done (Munshi and Dube, 1973; Munshi *et al.*, 1978, 1982; Ojha *et al.*, 1977, 1978; Ghosh and Munshi, 1987 and Ghosh *et al.*, 1990 etc.). Observation on certain Indian major carps have been made by Roy and Munshi (1984), Kunwar *et al.* (1989), Das *et al.* (2003) etc. However, detailed information on the relationship between oxygen uptake and body weight in the early developmental stages of *Catla catla* and *Labeo rohita* are not available. Although Motwani and Bose (1957), Saha (1966) and Singh (1977) etc. have studied the rate of oxygen uptake in the early developmental stages of certain Indian major carps, yet they have not established any relationship between oxygen uptake and body weight in the larval fishes. The present work is an endeavour to establish the relationship between the oxygen uptake and body weight in the early developmental stages (larval forms) of two Indian major carps, *Catla catla* and *Labeo rohita*. Necessities for such studies are felt for the purpose of transport and stocking of the larval stages.

Materials and Methods

The spawn and early fry of *Labeo rohita* and *Catla catla* were procured from a commercial fish farm of West Bengal and were transported to the Post-Graduate Department of Zoology in aluminium hundies (a large mouthed container). They were maintained in large glass aquaria with chlorine free water and constant aeration facilities. When the yolk sac was absorbed they were fed with artificial food and plankton. The early developmental stages used in the present study are spawn and early fry (Chakraborty and Murty, 1972) but the term larval fish is used here collectively for both the stages. The fishes were starved for 24 hours before experimentation.

A cylindrical respirometer similar to that used by Munshi and Dube (1973) was used with minor modification. A 100 ml. plastic syringe was used in place of a cylindrical glass respirometer (Plate 8, Figure 1). This was done to achieve better results with small sized fishes. The respirometer was connected to a constant level water tank. As determination of oxygen uptake of a single specimen was not possible, the specimens were taken in a group of 15 to 20. The flow of water through the respirometer was adjusted according to the size and number of specimen. The fishes were acclimatized to the respirometer for about two hours before experimentation. To avoid visual disturbances, the respirometer was covered with a black cloth during experimentation.

Inspired and expired water samples were collected in conical flasks and the dissolved oxygen content was measured using Winkler's volumetric method (Welch, 1948). The specimen from respirometer were removed after experimentation. Only the live specimens were dried on a blotting paper and weighed to the nearest milligram while the dead specimen were discarded.

Oxygen uptake was determined by the difference in the level of dissolved oxygen between inspired and expired water, the rate of water flow per unit time and the average weight of the fish.

The available data were analysed by linear logarithmic transformations using least square regression method. Log/log graphs were also plotted to show the relationship between oxygen uptake and body weight (Plate 9, Figure 1 and 2). Tables 4 and 5 incorporate the data on oxygen uptake (O_2) in *Catla catla* and *Labeo rohita* respectively. The values for regression coefficient (b), intercept (a) and correlation coefficient (r) for both the species are shown in Table 6.

Result and Discussion

(a) Oxygen Uptake per Unit Time ($\mu lO_2.h^{-1}$)

The oxygen uptake (O_2) increased with the increases in body weight in both the species. In *Catla catla* the oxygen uptake increased from 60.736 to 528.324 $\mu lO_2.h^{-1}$ with an increase in body weight from 15.0 mg to 1000 mg (Table 4). Similarly the

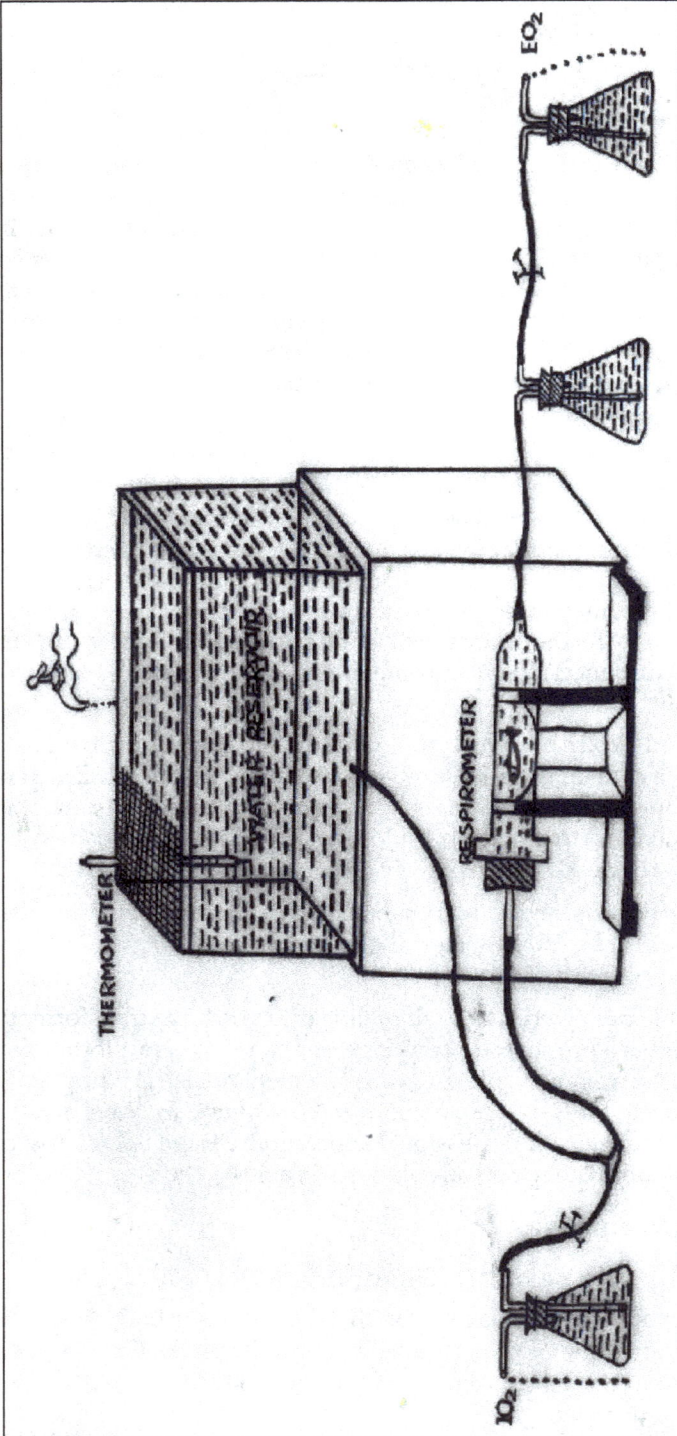

Plate 8

Figure 1: Experimental Set-Up

oxygen uptake in *Labeo rohita* increased from 20.353 to 314.225 $\mu lO_2.h^{-1}$ with an increase in body weight from 5.0 to 903.0 mg (Table 5).

Table 4: Oxygen Uptake (O_2) in Relation to Body Weight in Early Developmental Stages of *Catla catla* at 29.5±1°C

Body Weight (mg)	Oxygen Uptake (O_2)	
	$\mu lO_2.h^{-1}$	$\mu lO_2.mg^{-1}.h^{-1}$
15.0±3.00	60.736±2.199	4.141
70.0±2.52	98.401±4.598	1.406
80.0±2.61	104.041±2.001	1.301
140.0±0.00	148.844±0.000	1.063
150.0±0.00	153.476±0.000	1.023
230.0±3.51	245.659±3.632	1.068
460.0±4.04	361.791±12.913	0.787
630.5±11.02	414.783±16.352	0.658
840.0±19.80	457.139±7.573	0.544
1000.0±10.00	528.324±38.441	0.528

Table 5: Oxygen Uptake (O_2) in Relation to Body Weight in Early Developmental Stages of *Labeo rohita* at 29.5±1°C

Body Weight (mg)	Oxygen Uptake (O_2)	
	$\mu lO_2.h^{-1}$	$\mu lO_2.mg^{-1}.h^{-1}$
5.0±0.82	20.353±0.956	4.128
10.0±1.00	30.041±2.099	3.007
25.0±0.00	48.635±0.000	1.945
75.0±3.00	96.789±3.033	1.291
124.7±2.52	116.738±4.121	0.936
150.0±3.60	130.089±6.576	0.867
248.3±7.64	213.503±3.509	0.859
306.0±6.00	226.961±2.131	0.742
555.0±5.00	238.941±3.874	0.430
625.0±12.36	263.629±7.457	0.422
903.0±4.04	314.225±1.987	0.343

The log/log plot of oxygen uptake per unit time in relation to body weight gave a straight line with slopes of 0.5612 in *Catla catla* and 0.5384 in *L. rohita*. A high degree of correlation ($r = 0.9839$; $p < 0.001$) in *Catla catla* and ($r = 0.9939$; $p < 0.001$) in *Labeo rohita* have been found between the body weight and oxygen uptake (Table 6 and Plate 9, Figure 1 and 2).

The estimated value of a 1,10100 and 1000 mg *C. catla* comes to 10.464, 38.009, 138.714 and 505.037 $\mu IO_2.h^{-1}$. Similarly in *L. rohita* it was calculated to be 8.889, 30.746, 106.222 and 366.979 $\mu IO_2.h^{-1}$ respectively (Table 6).

**Table 6: Intercept 'a' Regression 'b' and Correlation Coefficient 'r'
to Show the Relationship Between Oxygen Uptake
O_2 ($\mu IO_2.h^{-1}$ and $\mu IO_2.mg^{-1}.h^{-1}$) and Body Weight**

Fish Species	Body Weight vs. Oxygen Uptake	Intercept 'a'	Slope 'b'	Correlation Coefficient 'r'
Catla catla	O_2 ($\mu IO_2.h^{-1}$)	10.4644	0.5612	0.9839P<0.001
	O_2 ($\mu IO_2.mg^{-1}.h^{-1}$)	10.7019	-0.4426	-0.9729P<0.001
Labeo rohita	O_2 ($\mu IO_2.h^{-1}$)	8.8994	0.5384	0.9989P<0.001
	O_2 ($\mu IO_2.mg^{-1}.h^{-1}$)	8.9778	-0.4632	-0.9919P<0.001

(b) Oxygen Uptake per Unit Body Weight ($\mu IO_2.mg^{-1}.h^{-1}$)

The oxygen uptake per unit body weight ($\mu IO_2.mg^{-1}.h^{-1}$) decreased with increasing body weight. With an increase in body weight from 15.0 to 1000.0 mg the rate of oxygen uptake decreased from 4.141 to 0.528 $\mu IO_2.mg^{-1}.h^{-1}$ in *Catla catla* (Table 1). Similarly it decreased from 4.128 to 0.343 $\mu IO_2.mg^{-1}.h^{-1}$ in *Labeo rohita* with an increase in body weight from 5.0 to 903.0 mg (Table 5).

The log/log plots of oxygen uptake per unit body weight gave straight lines with slopes of -0.4426 and -0.4632 respectively (Figure 6) for *C. catla* and *L. rohita*. The two variables show negative correlation both in *Catla catla* (r = -0.9789; p<0.001) and in *Labeo rohita* (r = -0.9919; p<0.001) (Table 6).

Some information on the oxygen uptake in early developmental stages of certain Indian major carps are available from the works of Motwani and Bose (1952), Saha (1966) and Singh (1977) etc. The oxygen uptake of 15 *Labeo rohita* (total weight 1.9 to 1.941 g) was reported to be 0.975 to 1.113 mg$^{-1}.h^{-1}$ by Motwani and Bose (1957). Saha (1966) reported that 50 fries of *Catla catla* and *Cirrhinus mrigala* consume 0.6 $mgO_2.h^{-1}$, which was higher than the rate of oxygen uptake by the same number of fries of *Labeo rohita* which consumes only 0.04 $mgO_2.h^{-1}$. Singh (1977) reported a higher rate of oxygen uptake 0.271 $mgO_2.h^{-1}$ in the fingerlings of *Labeo rohita* as compared to 0.179 $mgO_2.h^{-1}$ in the fingerlings of *Cirrhinus mrigala*. Recently Das *et al.* (2003) have reported that the fingerlings of *L. rohita* consumes maximum oxygen (169.87 $mgKg^{-1}h^{-1}$) per unit body weight followed by *C. mrigala* (155.52 mg kg$^{-1}.h^{-1}$) and *C. catla* (136.60 mg kg$^{-1}.h^{-1}$) during the initial six hours of study in a water having low dissolved oxygen.

The present study on the oxygen uptake in relation to body weight in the larval stages (early developmental stages) of *Catla catla* and *Labeo rohita* shows an interesting relationship. With an unit increase in body weight the oxygen uptake increased by a power of 0.5612 in *Catla catla* while in *Labeo rohita* it increased by a power of 0.5384. The weight specific oxygen uptake decreased by a power of -0.4427 in *Catla catla* and -0.4631 in *Labeo rohita*. The intercept value for *Catla catla* was calculated to be

10.702µlO$_2$.mg^{-1}.h^{-1} (= 10702 mlO$_2$.Kg^{-1}.h^{-1}), while for *Labeo rohita* it was 8.978 µlO$_2$.mg^{-1}.h^{-1} (= 8978 mlO$_2$.Kg^{-1}.h^{-1}) (Table 3). These results indicated that even in the larval stages the oxygen uptake is dependent on body weight as well as the fish species. Different slope values for different water breathing teleosts have been suggested by various workers. In *Cirrhinus mrigala* two different slope values have been reported by Roy and Munshi (1984) for summer (b = 0.8113) and winter (b=0.7961) with an increase in body weight from 5.0 to 180.0 g. A regression coefficient of 0.7469 and 0.8515 for *Catla catla* and *Labeo rohita* respectively has been reported by Kunwar *et al.* (1989) in fish ranging between 6.5 to 425.0 g. The above reported values are higher than the regression values obtained in the present study for *Catla catla* (0.5612) and *Labeo rohita* (0.5384) in the body weight range of 5.0 to 1000.0 mg. Thus, differences appear to be the effect of different life cycle stages, activity, change of feeding behaviour of the fish from yolk to plankton.

Kamler (1976) in *Cyprinus carpio* reported a slope value of 0.97 for fishes upto 674.0 mg and 0.80 for fishes weighing between 1.2 to 46.0g. Such intraspecific variation

Plate 9

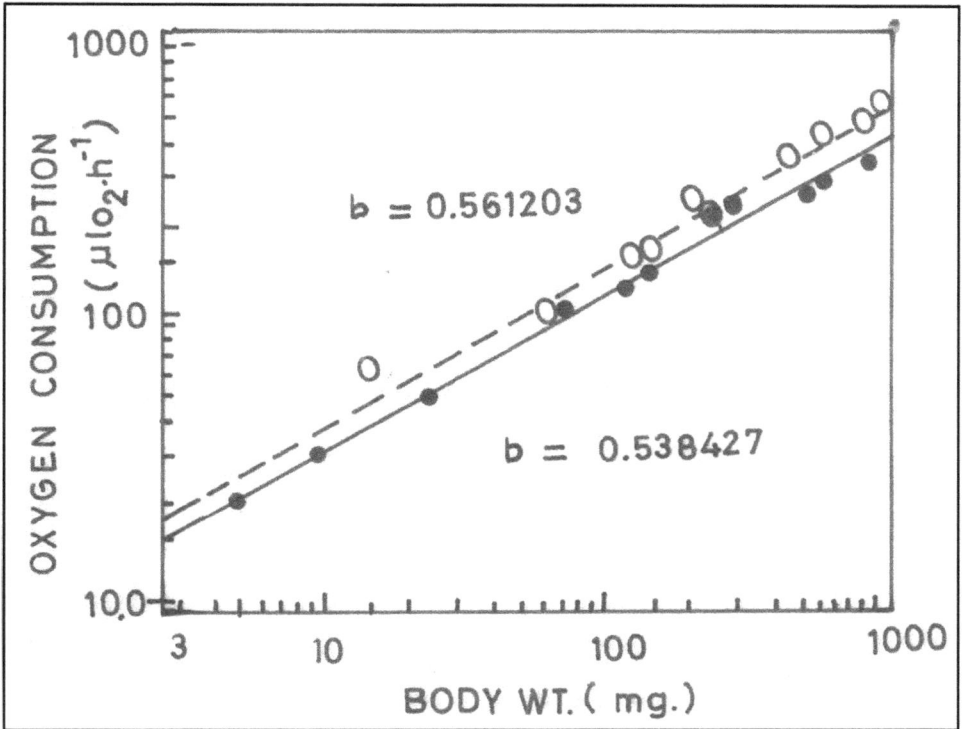

●: *Labeo rohita*, O: *Catla catla*

Figure 1: Log/Log Graph Showing Relationship Between Body Weight and Oxygen Consumption per Unit Fish (mlO$_2$. h^{-1}) in the Early Stages of *Catla catla* and *Labeo rohita*

Plate 9

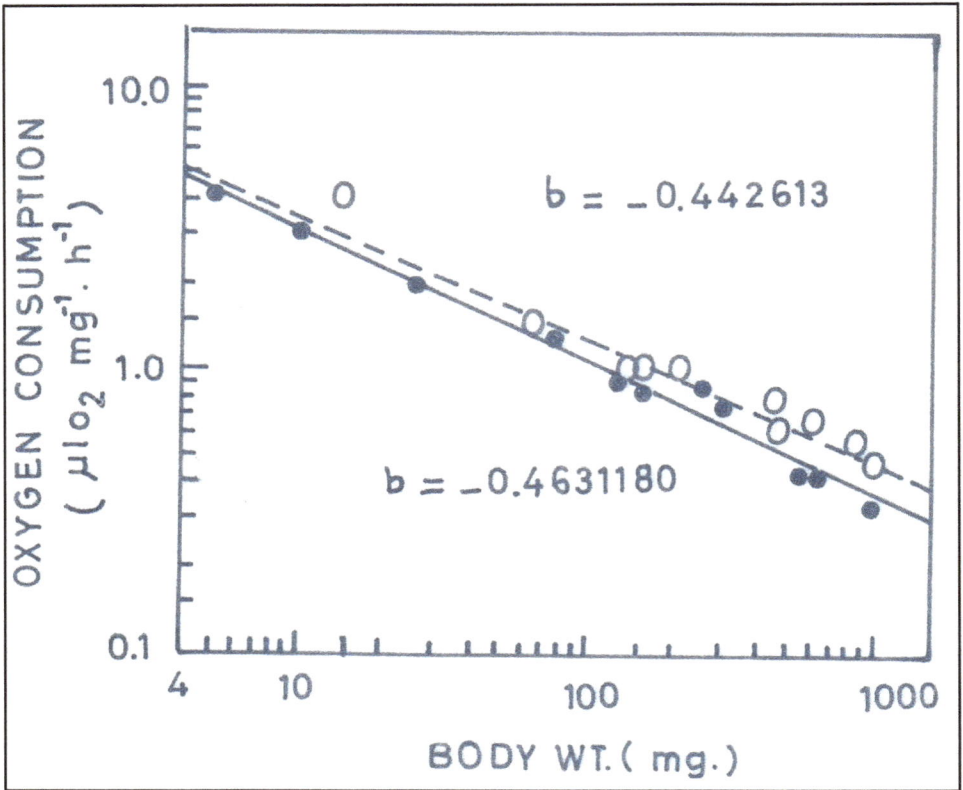

●: *Labeo rohita*, O: *Catla catla*

Figure 2: Log/Log Graph Showing Relationship Between Body Weight and Oxygen Consumption per Unit Body Weight (mlO$_2$. g^{-1}.h^{-1}) in the Early Stages of *Catla catla* and *Labeo rohita*

in the rate of oxygen uptake in an air-breathing fish *Heteropneustes fossilis* has been reported by Munshi *et al.* (1978). They have reported a regression coefficient of 0.478 for the juveniles while for adults it was 0.799.

Different slope values from 0.65 to 1.69 have been reported for different larval fishes by various workers *viz.* Desilva and Tytler (1973), Laurence (1978), Almatar (1984) and Giguere *et al.* (1988). After reviewing the data Giguere *et al.* (1988) concluded that standard and routine metabolic rate scale isometrically as $W^{1.0}$ in larval fishes but allometrically as $W^{0.8}$ in juveniles and adult fishes. But, the present findings contradict the generalization. The slope values for the larval fishes are lower than the larger fishes. Similar contradictions have also been reported by Pandey (1988). She reported two slope values for oxygen uptake in two weight groups of *L. rohita* i.e., 0.5976 for larval fishes (body weight 5.0 to 260.0 mg) and 0.8576 for fishes of higher weight group (16.0 to 425.0 g.).

The present study shows that the estimated value of oxygen uptake for a 1.0 g larval *Catla catla* is 0.505 mlO$_2$.h^{-1} which is higher than the oxygen uptake by a 1.0 g of *Catla* (calculated on allometric equations for fishes of higher weight group) which consumes only 0.420 mlO$_2$.h^{-1} (Kunwar *et al.*, 1989). Similarly, a 1.0g larval *Labeo rohita* has been found to consume 0.367 mlO$_2$.h^{-1} while *Labeo rohita* of 1.0 g (calculated from a higher weight group fish) consumes only 0.321 mlO$_2$.h^{-1} (Kunwar *et al.*, 1989). Thus, it is evident that the rate of oxygen uptake in the larva is higher in comparison to the fishes of higher weight group. It also suggests that the inflexion of slope values is perhaps associated with change in the feeding habit or the growth of the respiratory organs. Bennett (1988) discussed three determinants for metabolic rate *i.e.* temperature, body mass and phylogeny. However, the morphological system for oxygen extraction is designed in such a way that it can meet not only the determinants of metabolism but also the physiological, developmental and ecological constraints.

As the slope values are lower than 1.0, weight specific oxygen uptake decreased by a power of 0.4632 in *Labeo rohita* and 0.4426 in *Catla catla*. Similar negative regression coefficients are also reported by different authors (Roy and Munshi, 1984; Kunwar *et al.*, 1989 and Singh *et al.*, 1991). Prigogine and Wiame (1946) said that metabolic rate can never increase but only decrease with evolution in body size in organisms. There have been several different classes of explanations *viz.* structural constrains, opportunities for lower costs based on structural changes with size, changes in skeletal mass, a metabolically inert constituent, changes in energy devoted to growth with size (Schmidt–Nielsen, 1984). But, the explanations of structural constraints and opportunities for lower costs seem to be the main cause for the allometric growth of metabolic rate. Bennett (1988) emphasized that allometric dependence of metabolic rate resides in the structural and functional organization of the cells.

It has been reported that the juvenile fishes have more haemoglobin and erythrocyte number than adults (Dube and Munshi, 1973; Pandey *et al.*, 1976 and Mishra *et al.*, 1984), which helps in the extraction of more oxygen. Oikawa and Itazawa (1985) have observed that the gill area per unit body weight in larval *Cyprinus carpio* is more than the fishes of higher weight group. All these, support the present findings that larval fishes take up more oxygen than fishes of larger weight group.

When oxygen uptake of two carps are compared, it has been found that *Catla catla* consumes more oxygen than the *Labeo rohita*. The gill surface of 1.0 g *Catla catla* is higher (1505.22 mm^2/g) (Kunwar, 1984) than the *Labeo rohita* (873.73 mm^2/g) of same body weight (Pandey, 1988). Gill area of fish is considered to be directly proportional to the oxygen uptake. More oxygen uptake of *Catla catla* indicates its higher metabolic activity than the *Labeo rohita*.

Chapter 14

Fish Seed Packing and Transport

Transport of fish seed *viz.* spawn, fry and fingerlings of all cultivable species of fishes for stocking on culturable water has always been a problem for a long time ago. A large scale mortality of fish seed during transport is very common. Although a lot of attention has been given to improve transport techniques, still more is due to meet ever increasing demand so as to achieve high fish production in the country through intensive or high-tech aquaculture. The success of fish culture is to a good extent dependent on safe transport of fish seed. As satisfactory results are achieved, the methods on safe transport are largely empirical. The technique of transporting fish seed in a sealed carrier under the in closed atmosphere of oxygen is a break through in this field. Still there is lot of scope for further improvement in the current transport techniques. The name and size of the fishes in case of Indian major carps are as follow:

Spawn	⟶	Hatchlings measuring upto 5 mm in length.
Hatchling	⟶	Hatchlings measuring between 5 to 15 mm in length
Early Fry	⟶	Fish seed measuring between 15 to 25 mm in length.
Fry	⟶	Fish seed measuring between 25 to 30 mm in length.
Advance Fry	⟶	Fish seed measuring between 30 to 40 mm in length.
Fingerlings	⟶	Above 40 mm and upto 150 mm in length.
Table size fish	⟶	Above 150 mm in length.
Fish seed	⟶	Spawn, fry and fingerlings.

The development of the technique of packing live fish seed in plastic bags with oxygen for long distance transport by road, rail and air has reduced the mortality to a considerable extent.

Causes of Mortality

A number of factors may be responsible for mortality of fishes during the process of transportation. They are as follows:

1. High carbon dioxide and deficiency of oxygen in water medium.
2. Toxicity due to accumulating waste like NH_3 and other metabolites.
3. Hyperactivity, strain and exertion of fish seed.
4. Infection contracted.
5. Physical injury
6. A sudden fluctuation of water temperature in the carriers.
7. Diseases.

These causes may be exclusive or contributory, hence remedial measures are mostly overlapping in their effects of course with varying emphasis.

Packing and Transport Methods

The techniques of packing and transport mainly depend upon the distance to be covered and the anticipated time of confinement of fish seed, their size, available facilities etc.

Open System

In this traditional system, the surface of the water carried with fish seed is kept open with or without circulation of water, air or oxygen. Traditional method of live carrier with fish seed is kept open with or without circulation of water, air or oxygen, traditional method of live fish seed transport followed in West Bengal is in the earthen aluminum hundies. The capacity of these hundies is generally limited to 23-35 liters but only 2/3rd of the total capacity *i.e.*, approximate 14 liters is filled with clear and cool water. The density of packing spawn (6-8 mm) is about 1000- 3000/l of water. The loaded carrier if they are to be manually transported are carried on bamboo slings and the man carrying the load moves with rhythmic jerky movement to keep the water well-aerated and to prevent the fish crowding at the surface. The fishermen add 58 gm pulverized red soil over the surface of the water on the assumption that is helped to settle down the suspended organic matter and dead fish and thus keep the water clean. The dead fishes are periodically removed with the help of a piece of cloth during long distance transport. The water is often partially changed. For transport of fingerlings, galvanized iron drum of 180 l or plastic pools of 250 l capacities are also used.

Closed System

In this system, the water surface is exposed to compressed air or pure oxygen introduced to fill the carriers, which are sealed air-tight. Carp seed is presently transported in India in an economical manner with negligible mortality over long distances involving considerable period of time, in polythene bags placed in empty 18 litres kerosene oil tins with a little water and oxygen under pressure. The polythene bag of 74 cm × 46 cm its thickness is 0.0625 cm, placed in an 18 litres empty kerosene tin with lid, 6-7 l of water is poured into the bag and fish seed is introduced. The bag is then passed down to the level of water to expel the air. Pure oxygen from a cylinder is then released slowly into the bag through a rubber pipe until it is blown up by the oxygen to the top of the tin container. The seed is ready for transportation after then.

Packing density of Indian major carp seed in plastic bags, as developed at the research organizations in India is 30,000 for a period exceeding 24 hours, 60,000 for period up to 12 hours when the seed are of 0.5 to 0.7 cm in size. For journey of 12 hours duration 5,500 seed can be packed when 1.0 cm long, 2,200 nos. when 2.0 cm, 600 nos. when 3.0 cm, 330 nos. when 4.0 cm, 225 nos. when 5.0 cm, 80 nos. when 6.0 cm, 70 nos. when 7.0 cm and 40 nos. when 8.0 cm in length. The packing density of 2.0, 3.0, 4.0 and 5.0 cm exotics carps are: 300-400, 200-300, 160-200, 100-150 numbers respectively for silver carp, 400-600, 300-400, 200-300, 150-200 numbers respectively for grass carp and 750-1000, 500-750, 300-500 and 200-300 numbers respectively for common carp.

Use of Drugs and Chemicals

The use of anesthetics and chemicals are gaining popularity in the effective transportation of big size live fish in India. Anesthetics reduce the metabolic activity, which in turn, results in reduction in oxygen consumption and Carbon dioxide production and excretion of nitrogenous wastes. Further, it controls the excitability of fish and thereby reduces chances of injury during handling.

The drugs and chemicals, used in connection with transport of live fish, are categorized as below. But prior to using drugs and chemicals for transport, the actual concentrations required should be known as tolerance limits.

Anesthetic

Chemicals used to anesthetize fishes include Tricaine methanesulphonate (MS_{222}), Chloral hydrate, Amyl alcohol, Tertiary amyl alcohol, Quinaldine, Sodium amital, Methyl paraphinol, Urathan, Ether etc. Various scientists recommended use of sodium amytal a barbiturate at the rate of 52-172 ppm, MS_{222} a very strong tranquilizer to arrest quick reflex activity in bigger fish at a dose of 1.6 ppm. Tertiary amyl alcohol at about 2 ml/4.5 litres. Chloral hydrate at a dose of 300 to 750 ppm. Amyl alcohol at the rate of 0.025 to 0.08 per cent.

Antiseptic and Antibiotic

Acriflavin at the rate of 10 ppm, Methylene blue at the rate of 2 ppm copper sulphate at the rate of 0.5 ppm, Potassium permanganate at the rate of 3 ppm, chloromycetin at the rate of 8-10 ppm and sodium chloride at the rate of 3 per cent are used. A bath, prior to packing of live fish with the above chemicals are recommended as prophylactic measure to guard against accidental introduction of parasites and diseases.

Buffer

Di-sodium hydrogen phosphate (Na_2HPO_4) at 1.5 g/litres added in transport carriers for fingerlings has been reported beneficial to counteract the acidity of water medium.

Absorbents

A few absorbents like activated charcoal, amber lit, pulverized earth and permutated have been used to remove CO_2 and NH_3 rapidly from the water medium.

Before releasing the brood fish seed in the pond at destination, they should be properly acclimatized to few environments. Temperature adjustment is done by dipping the bags or containers in the pond water for a few minutes and then gradually adding pond water in it.

Critical Problems and Constraints in Rising Fish Seed Production

At present we have about 2.35 million ha tanks and ponds and 2.35 million ha reservoirs in India. They require huge quantity of fish seed. Except in West Bengal and Andhra Pradesh, we are not able to produce adequate quantity of fish seed required for stocking in the tanks. Raising fish seed production is beset with several problems.

1. Technology Perfection

Six decades ago, when artificial fish breeding was not achieved, fish seed was collected from rivers and canals. It consisted of a mixture of desirable and undesirable species. Time, effort and money were wasted to segregate the desirable species from undesirable varieties. During 1957 when hypophysation for fish breeding was invented, it became easy to produce more seed of desirable species. H.C.G. and ovoprim are used for fish breeding, both assured results. In hatching fish eggs also, the conventional earthen pits and double hapas are replaced by modern carp hatching. Transport of fish seed in metallic open containers are replaced by oxygen packed polythene bags kept in metallic containers, so that the seed can withstand long distance transport for longer durations without mortality. Thus, there is advancement of technology for rising fish seed production.

2. Man Power Training

Fish seed production is a scientific and skilled job. It requires well trained man power. Several organizations are conducting man power training for raising fish seed production. Particularly it is a golden opportunity for educated unemployed youth to come forward and learn the technology, so that they can find self employment besides raising fish seed production.

3. Financing the Scheme

Fish seed farmers require initially huge financial investment. It is very difficult for an average citizen to invest so much money for constructing a fish seed farm. Nationalized banks shall come forward to finance the scheme with liberal loans. The bankers may be educated through work shops to convince them regarding the profitability of the scheme, along with case studies, so that they may not hesitate to finance the scheme.

4. Location of Site and Related Technical Problem

Fish seed farms shall be located in the place where there is adequate demand of seed. There shall be facility by road and rail from the near vicinity of the farm area, so that the seed can be transported easily to long distance.

The farm site itself shall be technically sound and productive *i.e.* the area in which the farm is constructed may not be sandy, gravelly or rocky but the site shall be clayed loam with a little slit or fine sand not exceeding 10–20 per cent. Adequate water facility shall be available. If the water can be let into the farm by granulation, it is most desirable. The water shall be slightly alkaline with total alkalinity in the range of 100-150 ppm and pH value of 7-8.

5. Fish Seed Trade and Concept of Fish Seed Banks

At present, there is no organized fish seed trade in any state. The entrepreneur looks for more margin of profit. The purchaser desires to purchase for lesser price. During peak seed production, in state like Bengal, there will be glut in production. At such time the prices fall down. If well organized seed banks are established, seed can be canalized to the states-places, where there is dearth of production. In the absence of seed banks, it is not possible to locate the place where there is buffer seed production. In this connection, we may site the example of fish seed syndicate of West Bengal. The seed banks, once established, shall stand up to expectation by supplying seed of required quality and quantity at a fixed price to the demanding organization so that there will be healthy growth of fish seed trade.

6. Social Upliftment Through Rising Fish Seed Production

In many parts of the country, many of the fishing community members live below poverty line. Middle men advance money to such community and purchase fish at throw away price. Thus, the poor fishermen are entangled in poverty. Some tribal sects back financial power to purchase fish seed, though they are fish eaters.

The government shall come forward to organize fishermen Co-operative and arrange liberal loan to save the fishermen from exploitation by middlemen. The Government shall stock fish seed free of cost in the tanks situated in tribal areas and allow the tribals to catch the fish. Constraints, if any shall be solved in this process, so that more seed is stocked in tanks and poorer strata of society are benefited by this method.

Chapter 15

Common Aquatic Weeds

Undesirable aquatic plants that utilize nutrients presents in the water bodies and are more harmful than the beneficial. It prevent effective utilization of water and reduces productivity.

On the basis of their habit and habitats, the aquatic weeds have been classified into following categories:

Type of Weeds

There are four types of weeds:

(a) Floating
(b) Submerged } Rooted to the bottom

(c) Emergent
(d) Marginal } Not rooted to the bottom

(a) Floating Weeds
These weeds have their foliage above the surface of water with roots hanging free under heath.

Examples: *Eichhornia, Lemna, Azolla* and *Wolffia* etc.

(b) Submerged Weed
This may or may not be rooted *i.e.* Devoid of roots.

Examples: *Ceratophyllum, Utricularia, Hydrilla, Najas, Ottelia, Vallisneria*, etc.

(c) **Emergent Weed**

They are rooted in the bottom but having their foliage are flowers above the water surface.

Examples: *Nymphaea* and *Nelumbe* etc.

(d) **Marginal Weeds**

They are mostly rooted and infest the shallow fore-shore areas of the water body.

Examples: *Typha, Ipomoea* and Cyperus.

Table 7: Aquatic Weeds with Key Identification Characters

Type of Weed	Botanical Name	Common Name	Field Identification Characters
(I) **Floating Weeds**	(*i*) *Eichhornia*	Water Hyacinth	Stem modified rhizome Leaves large with bladder-like petiole. Flower in bunches.
	(*ii*) *Pistia*	Water Lettuce	Stem tubular creeping. Roots fibrous. Leaves entire with silky texture.
	(*iii*) *Lemna*	Duck weed	Stem modified to flat frond. Single long root fibers. Stem branched. Roots filamentous. Leaves sessile.
	(*iv*) *Azolla*	Water velvet	Stem branched, slender, roots Mosquitoferm filamentous, leaves sessile, bilobate, flowers primitive spore.
(II) **Submerged weeds**			
(*a*) Rooted to the bottom	(*i*) *Hydrilla*	Hydrilla	Stem cylindrical. branched and solid. Roots hairy, leaves sessile and in groups.
	(*ii*) *Vallisneria*	Eel grass Tape weed	Leaves ribbon like. Short to 3 feet long and in cluster. Small flowers.
	(*iii*) *Chara*	Stone wort	Stem soft, simple or a thread like in whole with black spot, branched.
	(*iv*) *Najas*	Water Nymphs	Stem smooth, branched indefinitely with long internodes, roots filamentous, leaves small, sessile, pointed.
	(*v*) *Ottelia*	—	Stem rhizome, leaves round or ribbon like, 10-20 cm long petiole.
(III) **Not rooted to the bottom**	(*i*) *Ceratophyllum*	Horn whort	Born on 10-20 cm. long petiole. Stem with short internodes, roots absent, leaves 3 forked like.
	(*ii*) *Utricularia*	Draft bladder wort	Stem with short internodes. Hairy roots, usually filiform leaves, or whorls having tiny bladders.

Contd...

Eichhornia

Pistia

Lemna

Azolla

Hydrilla

Vallisneria

Chara

Nazas

Ottelia

Ceratophyllum

Utricularia

Nymphaea

Nelunbe (Lotus)

Marsuilea

Jussiaea

Plate 10

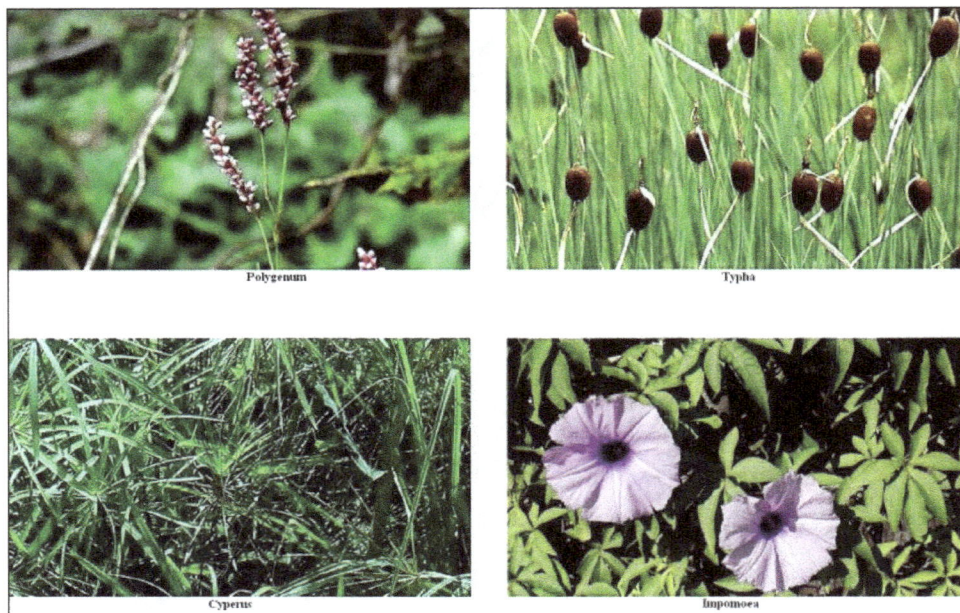

Plate 11

Table 7–Contd...

Type of Weed	Botanical Name	Common Name	Field Identification Characters
	(*iii*) *Nymphaea*	White water lilies	Creeping stem rhizome. Root tuberous and long. Disc like leaves, 2-2 feet in diameter. Notched borne on long petiole.
	(*iv*) *Nelumbe*	Lotus	Deeply rooted, saucer shaped leaves.
	(*v*) *Trapa*	—	Long stem and slender leaves long with inflated petiole.
(IV) **Marginal**	(*i*) *Marsuilea*	Water clover	Creeping stem or floating leaflets 3-5 with entire margins. Covering a long petiole.
	(*ii*) *Jussiaea*	Water purslane	Stem creeping, cylindrical branched. Roots thread like. Leaves sessile or petiolate.
	(*iii*) *Polygenum*	Water smart weed	Long stem, erect and branched.
	(*iv*) *Typha*	Cattails	Creeping stem rhizome. Upper part project out. 10-12 feet. Fleshy and triangular leaves.
	(*v*) *Cyperus*	Flat sedges	Triangular stem hard and erect. Long channelled leaves.
	(*vi*) *Impomoea*	Water spinach	Stem rounded, hollow and jointed, flowers big, purple-white-yellow in colour, leaves sessile.

Eradication

1. Choose the proper method for eradication of weeds, considering the local conditions.
2. Destroy the weeds before their active reproduction takes place.
3. Do not apply chemicals if weather conditions are not favourable for their effective utilization.
4. Ensure that the weeds collected for study do not enter the water body again.

Chapter 16
Common Aquatic Insects

Aquatic insects are less then 4 per cent of the total number of existing insects fauna. Nursery ponds designed to rear spawn of major carp and exotic carp but ponds are heavily populated during and after the season with aquatic insects. They prey directly upon carp spawn and fry also compete with later for food.

Identification

Phylum: Arthropoda

Class: Insecta

Order: Hemiptera

Order: Coleoptera

Order: Odonata

Examples

1. *Notonecta*
2. *Nepa* (Water Scorpions)
3. *Ranatra* (Stick insect)
4. *Belestoma* (Gaint water bug)

Table 8: Aquatic Insects

Phylum: Arthropoda

Class: Insecta

Order: Hemiptera

Family: *Nepidae* (Water Scorpions)

Genus: *Nepa*

Nepa

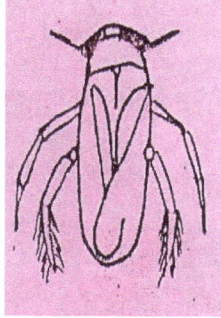

Corixa

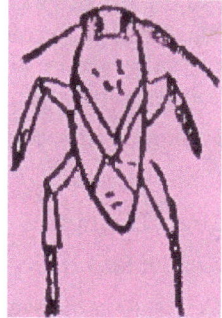

Notonecta

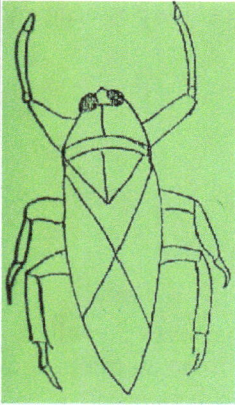

Belostoma

Cybister

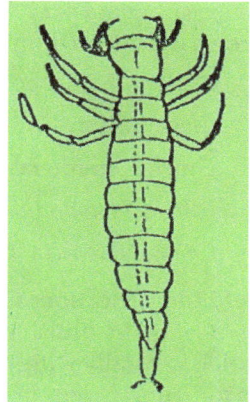

Cybister Larva

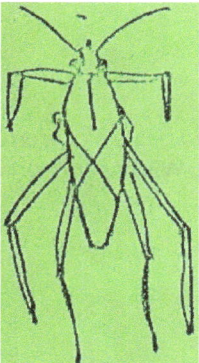

Gerris

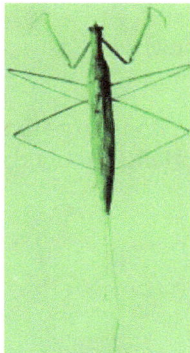

Ranatra

Dragonfly Nymph

Plate 12

Field Identifications Character

Blakish, dull in colour pray in a position head downward and forelegs held out in raptorial position. Rostrum short and strong. First pair of wings covers the body. Two long respiratory filament at the tip of the abdomen. Grows upto 17 mm in length.

Phylum: Arthropoda

Class: Insecta

Order: Hemiptera

Family: *Ranatridae*

Genus: *Ranatra* (water stick insects)

Field Identifications Character

Dul green in colour. Prothorax narrow than head. Forelegs long and modified for grasping the pray. Two long filamentous respiratory tubes at the posterior end. Grows upto 40-60 mm in length.

Phylum: Arthropoda

Class: Insecta

Order: Hemiptera

Family: *Notonectidae*

Genus: *Notonecta* (Back swimmer)

Field Identifications Character

Generally bluish black in colour. Swim jerkily body deep convex dorsally and keeled ventrally with side slopping hind legs long and fringed for swimming. Grow up to 10 mm in length.

Phylum: Arthropoda

Class: Insecta

Order: Hemiptera

Family: *Belostomatidae*

Genus: *Belostoma* (Giant water bug)

Field Identifications Character

Brown or dull greenish in colour. Body large flat and oval. Head bears a powerful beak. Forelegs chelatex and hind legs long modified for swimming. Wings functional. Grow up to 20-70mm.

Phylum: Arthropoda

Class: Insecta

Order: Hemiptera

Family: *Gerridae*

Genus: *Gerris* (water strider).

Field Identifications Character

Gregarious (live in group). Head small and body slender. Forelegs small while last two pairs of legs. 2-3 times as long as the body. Live on the surface film. Grows up to 2-15mm.

Phylum: Arthropoda

Class: Insecta

Order: Coleoptera

Family: Nytiscidae

Genus: Cybister (Diving beetle)

Field Identifications Character

Shining black or brownish black. Marked dull yellow margins. Head broad. Tightly fitting and capable of only slight movements. Antennae small and slender and hind legs long with hairs. Wings functional. Grows up to 35 mm.

Phylum: Arthropoda

Class: Insecta

Order: Odonata

Family: Nymphadae

Genus: Nymph (Dragon fly)

Field Identifications Character

Muddy brown in colour head brown and the lower side is concealed by the development of the lower lip in the form of long jointed arm like structure. Antennae short.

Chapter 17

Identification of Plankton

The community of mainly small organism both animals and plants that float or dip in water with limited or no power of locomotion is referred to as plankton. Plankton plays an important role in the aquatic environment. The fishes, prawns and other aquatic organisms are largely dependent on zooplankton for food during some part of their life cycle.

The word was first used by Victer Hensen in 1887. The individual organisms are called Plankters. Planktonic organisms could be classified in various ways:

1. According to mode of nutrition.
2.
   According to size.
3. According to habitat.
4. According to nature of life history.

On the basis of mode of nutrition it is divided in to:

(a) Phytoplankton's.
(b) Zooplankton's.

Phytoplankton's

Planktonic plants that have the power to synthesize their food from inorganic raw materials are found dissolved in water with the help of sunlight and photosynthetic pigments.

1. Class–*Cyanophyceae* (Blue green algae)
2. Class–*Chlorophyceae* (Green algae)
3. Class–*Bacilariophyceae* (Diatoms)

Cyanophyceae (Blue Green Algae)

These are conical or filamentous algae, which generally have a distinctive blue green colour, float near the surface. They have no well defined nucleus and have no chromatophores, very common in freshwater. Forming blooms in summer.

Examples: *Anabaena, Oscillatoria, Spirulina, Microcystis, Chrococcus* etc.

Chlorophyceae (Green Algae)

Chlorophyll is present in chloroplast, Unicellular filamentous and colonial in appearance.

Examples: *Volvox, Eudorina, Pandorina, Chlamydomonas, Pediastrum, Scenedemus, Spirogyra, Zygnena, Oedogonium and Euglena, Closterium, Cedogonium* etc.

Bacilariophyceae (Diatoms)

They are unicellular filamentous, conical, Choromatophores present, greenish yellow in colour. Silicoeus box like cells.

Examples: *Navicula, Synenda and Melosira.*

Zooplanktons

Planktonic animals are dependent on plants or other forms of particular organic matter for their food. They cannot prepare food from available inorganic matter.

1. Phylum Protozoa

Unicellular animals. These groups are less common in planktons of freshwater than in marine water planktons.

Ciliate

Paramecium is most important ciliates in freshwater.

Rotifera

They are multicellular animals with a ciliated corona at anterior and posterior end tapering into stalk.

Examples: Brachionus, Kellicottia etc.

2. Crustacea

Along with rotifers crustacean are most common group of animals among freshwater planktons.

Cladocera

Body enclosed in bivalve shall, head is out side with one pair of movable prominent eyes. Antennae on the head and used for swimming.

Examples: Daphania, Ceriodaphnia, Moina and Diaphanosoma etc.

Copepodes

Body divided into two sections. The metasome and tail or urosome 5 pairs of biomass legs. The copepods larvae are called, as Nauplius is also planktonic.

Calanoidae

Urosome is shorter than metasome, antennae long reaching to the end of urosome.

Examples: Cyclopodia, Cyclops etc.

According to Size

Macroplankton

Larger than 3 mm in size that could be seen by naked eyes.

Nanoplanktons

They are not retained by a net of 0.25 bolting silk. They are mostly phytoplankton.

Methods of Study of Plankton

Designed of Plankton Net

The net is of truncated cone shape. The upper diameter of the net, which receives the brass ring, is 30 cm the lower diameter of the net that receives the upper rim of the collecting bucket is 9.2 cm the side length is 84 cm the net may have canvas lining of 12.5 cm width at the top. The cloth is of 0.25 bolting silk.

Quantitative Estimation of Total Plankton

Settling Volume

Allow sufficient time about 24 hr or more for the plankton to settle in graduated cylinders and record its volume.

Quantitative Estimation of Plankters

Direct Census Methods
1. Take a microscope with suitable oculars and objections.
2. Take Sedgwick raffles (50 × 20 × 1 mm)
3. Note down the volume of the plankton concentrate.
4. Shake well and trendier one ml to the counting cell and cover it with a cover glass.
5. Count plankton's species wise or group wise in 10 square and find the average. Ten counting unit taken at random. It can be computed using this formula:

$$ni = \frac{(A\ 1000)\ C}{1}$$

ni: Number of planktons.

A: The mean number of plankters per counting unit of one mm^2.

C: Volume of concentrated plankton in ml.

l: Volume of original reservoir water in liter.

Plate 13: Some Common Zooplankton

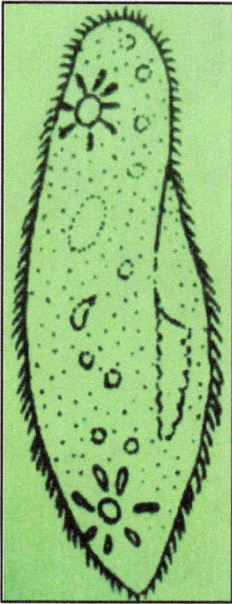

Paramecium

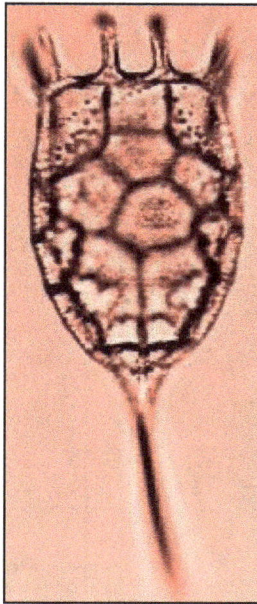

Keratella

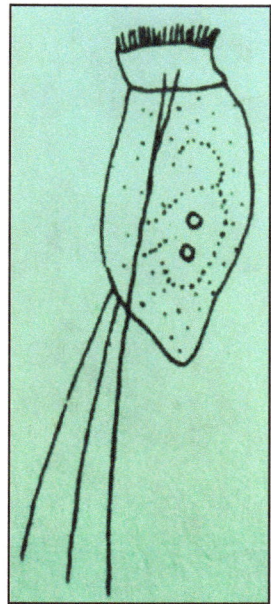

Filinia

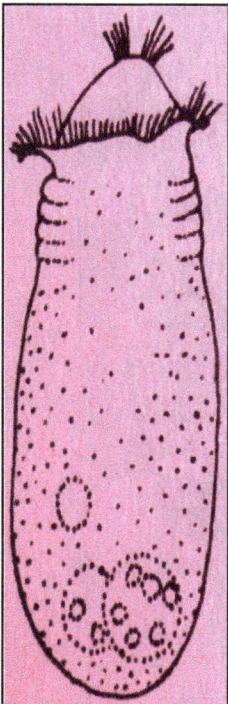

Asplanchna

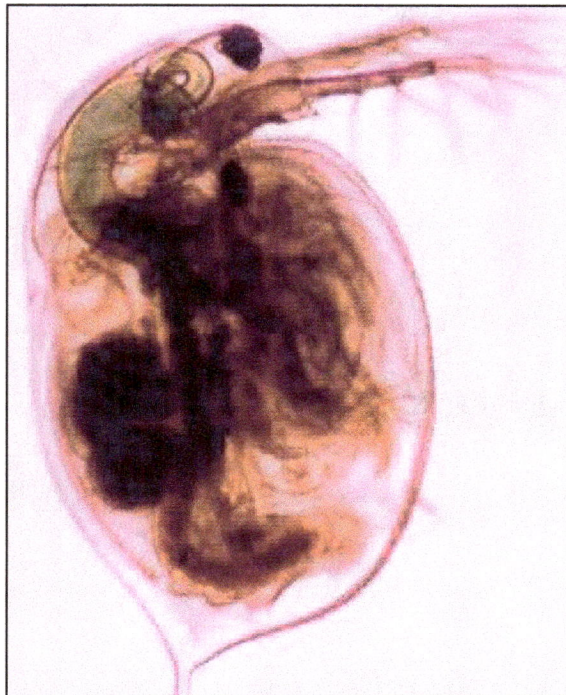

Daphnia

Contd...

Plate 13–Contd...

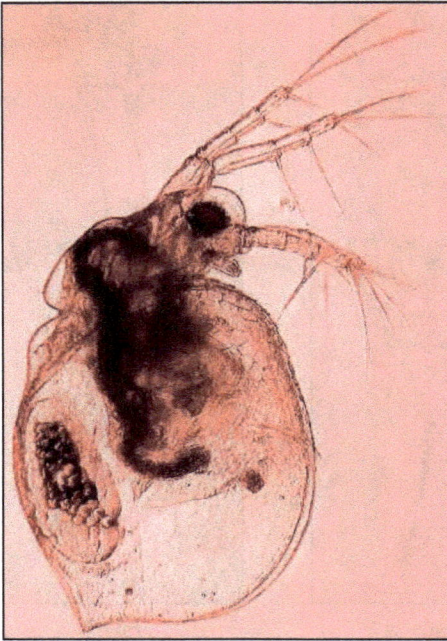

Ceriodaphnia

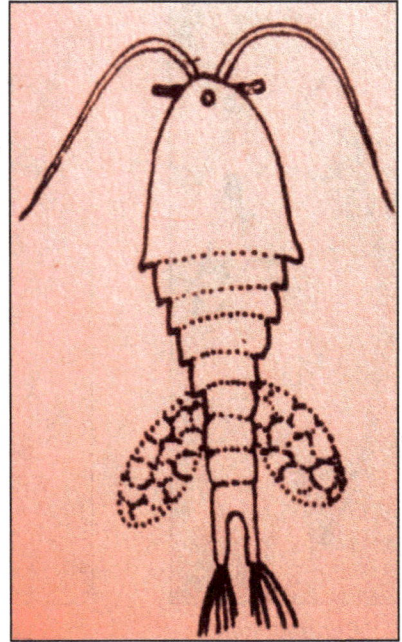

Cyclops

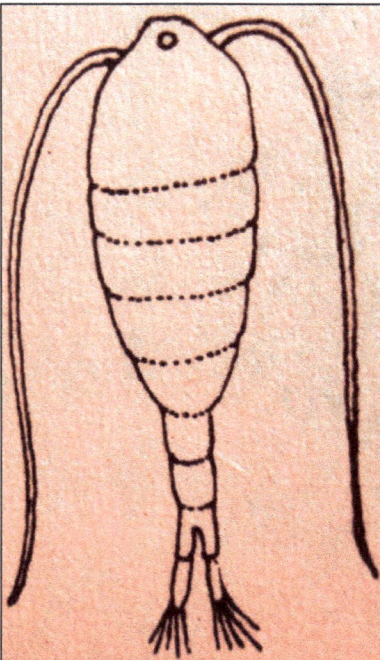

Diaptomus

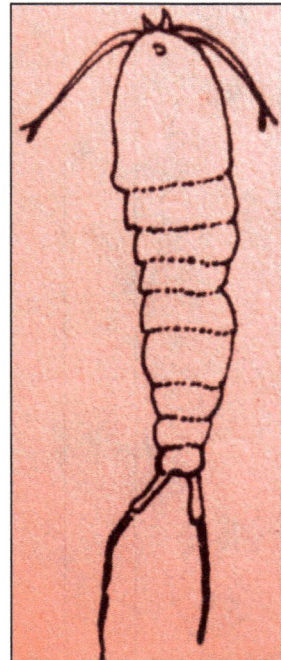

Canthocamptus

Contd...

Plate 13–Contd...

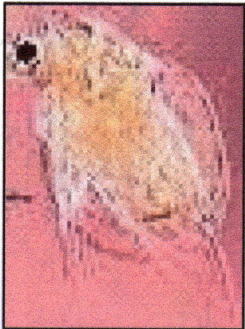

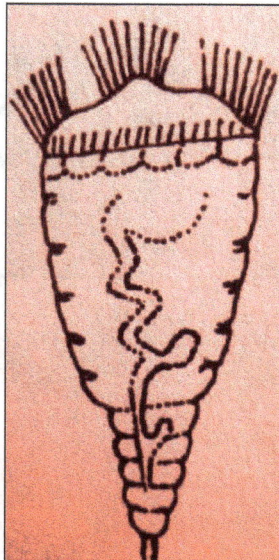

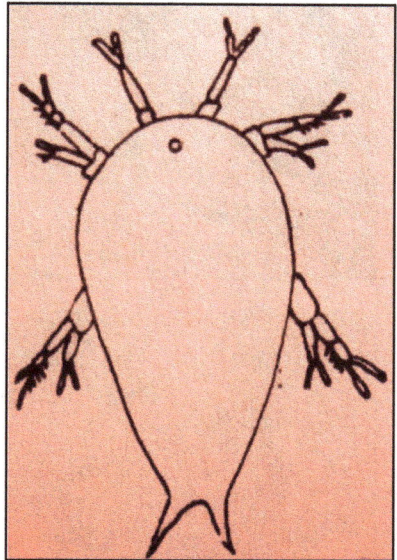

| **Monia** | **Brachionus** | **Naplius** |

For large plankters, however, can be counted through the entire counting cell of 1 ml capacity.

$$ni = \frac{WC}{1}$$

ni: The number of large plankters

W: The number of larger plankters in the entire counting cell of 1 ml capacity.

C: Volume of concentrated plankton in ml.

l: Volume of original reservoir water in liters.

Table 9: Field Identification of Common Zooplankton

Sl.No.	Class	Name of Plankton	Characters
1.	Ciliophora	Paramecium	Microscopic Unicellulor, organism body slipper shape covered with small cilia.
2.	Rotatoria	Keratella	Shape conical. 4-6 spines anterior and 1-2 posterior.
		Brachionus	Body enclosed in dumble shaped licica '6' of spines projecting anterior margin.
		Filinia	2-3 very long movable appendages extends from anterior side body shape indefinites and foot absent.
		Asplanchna	Body large sac–like. Lrica absent. Spines absent. Foot absent.

Contd...

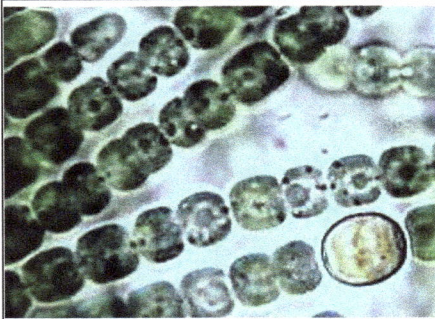
Anabaena

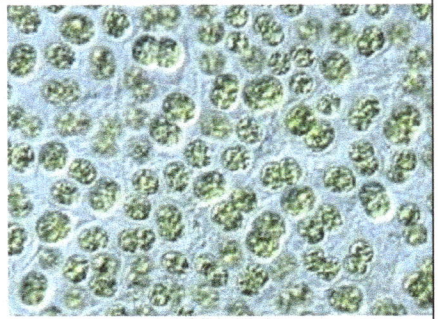
Microcystis

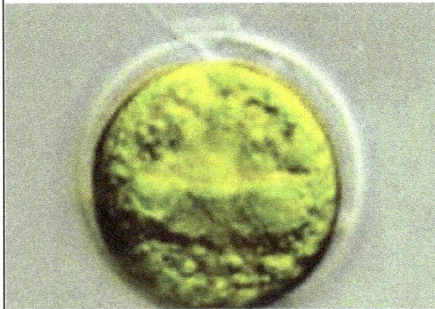
Chlamydomonas

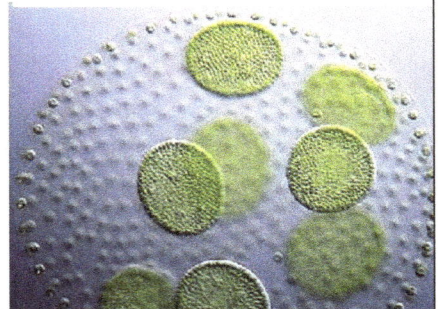
Volvox

Spirogyra

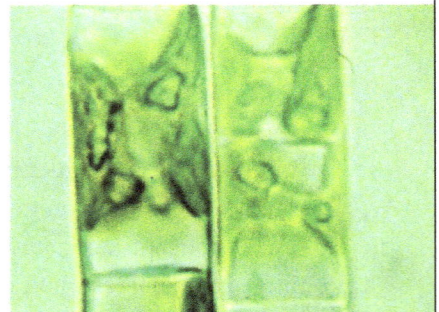
Ulothrix

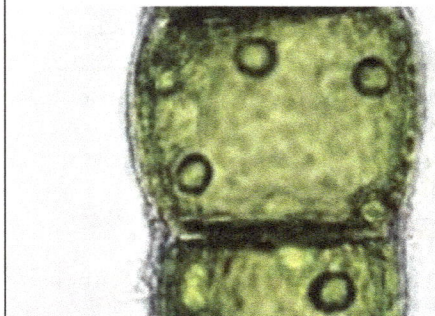
Oedogonium

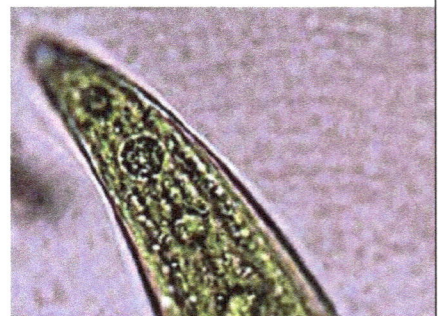
Closterium

Plate 14: Some Common Phytoplankton

Table 9—Contd...

Sl.No.	Class	Name of Plankton	Characters
3.	Crustacea (Cladorera)	Daphnia	Head keeled above. No transverse suture on neck. Shell with polygonal marks with post.
		Ceriodaphnia	Head without a beak. Small and depressed. Short dorsal spines.
		Moina	Heavy body with rounded abdomen. Head large. Antennules long and freely movable. Abdominal setae very long.
		Moinodaphnia	Valves elliptical. Head small. Antennules on ventral surface of head. Minute spiner on ventral margin sharp spine.
		Bosmia	Antennules large. Fixed and parallel 6 pairs of feed.
		Sida	Head large and separated from the body by a depression. Rostrum pointed and form a beak.

Table 10: Field Identification of Common Phytoplankton

Sl.No.	Class	Name of Plankton	Characters
1.	Mysophyceae	Nostoc	Colonial gelatinous thallus comprised of several interior wind and contorted moniform filaments having heterocyst or big cells in midway of each filament.
		Anabaena	Filaments small, single, like mostoc but never colonial.
		Microcystis	Unicellular, cells variable in shape. Often colonial.
2.	Chlorophyceae	Chlamydomonas	Microscopic, Unicellular, cell slender with blunt anterior end and pointed posteriorly. Single flagellum, chloroplast in lobules.
		Volvox	Rounded colonial form consisting of 500-60,000 cells. Marginal cells phyriform, while centre cells rounded in shape several flagella coming outside gelatinous mass.
		Spirogyra	Silky thread like filaments unbranched forming scam or mat each cell linearly elongated with ribbon-shaped chloroplast.
		Ulothrix	Filamentous, unbranched, cells brick shaped with single girdle shaped chloroplast.
		Oedogonium	Filamentous, unbranched, cells brick shaped linearly elongated reticulated chloroplast.
		Closterium	Microscopic, unicellular, bow-shape chloroplast many and star shaped.

Chapter 18

Estimation of Some Important Physico-chemical Parameters of Pond Water

Estimation of Dissolved Oxygen

Apparatres
1. 250 ml conical flask.
2. Measuring cylinder of 100 cc.
3. Pipette
    (*i*) 2 ml pipette (3 Nos.)
    (*ii*) 1 ml pipette (1 Nos.)
    (*iii*) Monopan balance.

Reagent
1. Alkaline iodide: Dissolve 50 g of KOH and 25 g of in 100 ml of boiled distilled water and filter.
2. Manganous Sulphate: Dissolve 50 g of $MnSO_4 4H_2O$ in 100 ml of boiled distilled water and filter.
3. Concentrated Sulphuric Acid. (sp.gr. 1.84)
4. Sodium thiosulphate, 0.025N: Dissolve 12.41 g
5. $Na_2S_2O_3 5H_2O$ in 500 ml of boiled distilled water and filter. Dilute it to 4 times to boiled distilled water to prepare 0.0025 N solution. Keep in a brown glass stopperd bottled.

6. Starch solution (as indicator): Dissolve 1 g of starch in 100 ml of warm water.

Procedure

Remove carefully the stopper of the sample bottle; add 2 ml of manganous sulphate reagent and 2 ml of alkaline Iodide reagent by means of two ml pipette dipped to the bottom of the bottle and slightly drawing out as the reagent are added. Replace the stopper and invert the bottle three or four times for a thorough mixing of the agents. A fluorescent precipitate will be formed which will settle at the bottom. If the precipitate is whiter in colour. Oxygen in very poor light to brown colour indicates poor oxygen while brown to red indicates medium to high oxygen. For quantitative estimation, add 1 ml of concentrated H_2SO_4 to dissolve the precipitate. Transfer 50 ml of this solution to a conical flask placed on white background and add $N/40\ Na_2S_2O_3$ drop by drop till the colour turns pale yellow, then 1 ml of starch solution is added to give blue colour and the titration is completed by turning it colourless.

Formula

$$\text{Dissolved oxygen, mg/l} = \frac{(\text{ml} \times N) \text{ of titrate} \times 8 \times 1000}{V_1 - V_2}$$

where,

V_1: Volume of sample bottle after placing the stopper.

V_2: Volume of the part of the contents titrated.

OR

Number of ml of $Na_2S_2O_3 \times 4 = $ ppm of dissolved oxygen.

Estimation of Free Carbondioxide

Apparatus

1. Conical flask 250 ml.
2. 10 cc pipette.
3. 1/10 cc pipette.

Reagents

1. Sodium hydroxide, 0.05N or N/44 NaOH: Prepare 1.0N NaOH by dissolving 40.0 g of NaOH in boil distilled water to make 1 litre of solution. Dilute 50 ml of 1.0 N NaOH to 1 litre.
2. Phenolphthalein indicator.

Procedure

Take 50 ml of the sample in a flask and add two drops of Phenolphthalein indicator. If the water remains colourless add 0.05 N NaOH drop by drop from 10 ml graduated pipette with a very gentle stirring with a glass rod till the colour turns pink.

Formula

$$\text{Free CO}_2\,\text{mg/l} = \frac{(\text{ml} \times \text{N}) \text{ of NaOH} \times 1000 \times 44}{\text{ml sample}}$$

OR

No. of or $N/44$ NaOH required $\times$ 20 ppm of free carbon dioxide.

Estimation of Total Alkalinity

Apparatus
1. Conical Flask (2 nos) of 250 ml.
2. Beaker 500 ml.
3. 10 cc pipette.

Reagents
1. HCl, 0.1N: Dilute 12n concentrated HCl (sp. Gr. 1.18) to 12 times (8.34 to 100ml) to prepare 1.0 n HCl. Dilute it further to make 0.1 N HCl (100 to 1000 ml).
2. Phenolphthalein indicator.
3. Methyl orange-indicator.
4. Sodium carbonate, 0.1 N: dissolve 5.3 g of Na_2CO_3 in distilled water to prepare 1 litre of solution.

Procedure
Take 100 ml of the sample in a conical flask placed over while porcelain tile. Add two drops of Phenolphthalein indicator. If the sample remains colourless, Phenolphthalein is absent. If the sample turns pink titrate it with 0.1 N HCl from a pipette up to a colourless end point. This is Phenolphthalein alkalinity (PA).

Methylorange Alkality (MP)
Proceed in some way as before using methyl orange as indicator, the end point being indicated by a colour change from yellowish faint orange colour. Note down the total Number of ml 0.1 N HCl required for complete titration.

This is total alkalinity.

Formula

$$\text{PA as CaCO}_3,\,\text{mg/l} = \frac{(\text{A} \times \text{Normality}) \text{ of HCl} \times 1000 \times 50}{\text{ml of sample}}$$

$$\text{TA as CaCO}_3,\,\text{mg/l} = \frac{(\text{B} \times \text{Normality}) \text{ of} \times 1000 \times 50}{\text{ml of sample}}$$

OR

Number of ml $N/50$ H_2SO_4 required $\times$ 20 = ppm of total alkalinity.

where,

        A: ml of HCl used with only Phenolphthalein.

        B: ml of total HCl used with only Phenolphthalein and methyl orange.

    PA: Phenolphthalein

    TA: Total alkalinity

Estimation of pH

1.   With the Help of Portable pH Meter

Reagents

Buffers of different pH

Follow the instruction given by the manufacture to use the pH meter. Use the Phosphate buffers for calibration. After that the pH of unknown water is detected by the pH meter. For field work the pH paper of different pH is used for rough diction.

2.   With the Help of pH Paper

pH paper of both broad and close ranges are available. Those are dispensed in reels. The colour of the paper changes with pH. The colours produced in contact with solutions of different pH are given on the reel for composition.

The broad range covers 1 to 5, 4 to 6, 6 to 8, 8 to 10, 10 to 12, 12 to 14.

The narrow range covers fractions of each pH.

Determination of Turbidity

Apparatus

    Secchi Disc.

Procedure

A standard secchi disc consists of a circular metal plate of 20 cm diameter. The surface of which is divided into 4 equal gradients painted black and white alternatively with lower side black to eliminate reflection of light from the side. The disc is lowered into the water and depth (d_1) at which the disappear is noted. Now the disc is lifted slowly and the depth (d_2) at which the disc reappears is noted. The feeding (d_1+d_2) in cm gives a measure of light penetration on is known as Secchi Disc transparency.

Chapter 19

Common Fish Diseases and their Control

Diseases can be defined in a simple way as "any sort of disorder or the abnormal activity of living organism leading to its unfitness". Studies related to fish diseases are gaining due importance now-a-days with large scale increase in fishery management and development programme.

The symptoms of diseases in fish differ according to their nature, such as unusual movement, unhealthy look, discolouration, a film covering on the skin, blisters, swelling or spotting etc. disease of the skin are cauterized by formation of slimy, great excretion, covering small or greater areas with white brownish or even black spots. The gills of the fish will have a reddish colour, whereas pale gills are the sign of a disease. Presences of white or red spots on the gills are the symptoms of the disease.

Like all other animals, fishes are also prone to several parasitic and non-parasitic diseases. The outbreak of any disease is preceded by adverse hydrological conditions during summer when the water levels are low. The bruises on the body of fishes are found as a result of mechanical injuries during fishing or transport facilitates parasitic infections. The intensity of stocking is also another factor for parasitic infection. If the diet of fish is inadequate the fish will grow weak and exposed to infection.

The fish diseases are important for three main reasons. They are:

1. The number of fish available for consumption is reduced.
2. The palatability of diseased fish may be reduced, and
3. The danger of fish parasites occurring in food, which may threaten human health. Based on causative points, the fish diseases may be classified as follows:

    A. Parasitic infectious diseases, and

    B. Non-Parasitic Affections.

A. Parasitic Infectious Diseases

I. Bacterial Diseases

Bacterial diseases are responsible for heavy mortality in both wild and cultured fish. Infection may occur in the internal organs like muscles, skin and also in fins. Whenever skin or muscles or both are affected, red spots are developed. The different types of diseases commonly caused by bacteria are as given below:

1. Fin and Tail rot
2. Ulcer Disease
3. Dropsy
4. Eye Disease
5. Tuberculosis
6. Furunculosis and
7. Vibriosis.

Fin and Tail Rot

The causative agent of this disease is bacteria. They cause considerable loss to the fishes. Adult and young fishes are affected by this disease. As a result of this disease, tail or fins or both get putrefied. Tail becomes torn and is gradually consumed by bacteria.

Control

Dip treatment with 500 ppm $KMnO_4$, improve water quality, reduce organic load.

It can be controlled by bath treatment in 1:20,000 solutions of $CuSO_4$ till fish shows distress or 10 to 15 minutes of infection portion.

Ulcer Disease

This disease is generally observed in Major Carps, Grass Carp and in some air-breathing fishes. The pathogenic bacteria are responsible. The disease is caused by *Pseudomonas punctata*.

Symptoms

On the body of the fish, open sores which are called as Ulcers are formed. These increases in size gradually and exposing the muscles. Ulceration, exopthalmia and abdominal distension.

Control

Dip treatment for one minute in 1:2000 $CuSO_4$ for 3–4 days. Seriously affected fishes should be destroyed to further spread of the disease. As a preventive measure, $KMnO_4$ at the @ 0.5 ppm should be used as disinfectant onto the pond. Oral feeding of Oxytetracycline, Terramycin and Chlorompheniol @ 75-80 mg/kg to feed for 12 days.

Dropsy

This is a very serious disease affecting the major carps which is caused by *Pseudomonas punctata*.

Symptoms

Internal organ gets accumulated with liquid and water. This disease affects the belly which may swell to a considerable extent so that the fish belly looks like a balloon shaped which is about to burst. However, fish dies before pressure become great enough for this to happen.

In most of the cases intestines were highly inflamated and the liver is badly affected. Scale protrusion due to internal pressure and also inflammation of eye (Exopthalmic).

Control

1. Dip treatment in 1 ppm $KMnO_4$ for 2 minute.
2. Antibiotics like chloromycetin is used @ 60 mg/kg body weight.
3. During treatment, no food should be given.

Eye Disease

This is an epidemic disease and generally found in medium sized and large scale *Catla catla*.

Causative Agent

Aeromonas liquifaciens.

Symptoms

During the initial stages, Cornea of the eye becomes vascularised and later become opaque; subsequently eyeball gets petrified leading to death.

Control

1. Disinfection of pond with 0.1 ppm of $KMnO_4$ followed by 300 ppm lime.
2. Chloromycetin 8-10 mg/l bath for 1 hr. for 2–3 days.
3. Maintain oxygen level.

Tuberculosis

Some of the freshwater and marine water fishes are reported to be infected with Tuberculosis bacteria.

Symptoms

1 Lack of apetite, progressive thinness and sluggish movement.
2. Affected fish becomes dark and show swelling in the abdomen.

Internal Manifestation

Tuberculosis may be found in any organs but especially in Liver, Spleen and Kidney.

Control and Treatment

No treatments are known and affected fish stock should be destroyed. No restocking should be done before disinfecting the pond.

Furunculosis

Furunculosis is characterized by formation of boils that are seen as round swellings on the sides of the body.

Causative Agent

Aeromonas salmonicida.

Symptoms

First symptoms occur in internal organs–intestines become heavily implimated and turn blood red colour. Fish may be out at this stage. Later on boils over the body are also seen.

Control

1. Removal of all fishes showing furuncles.

2. Treatment with antibiotics and sulphonamides.

Vibriosis

Normally it occurs in brackish water fishes when temperature remains high.

Causative Agents

Vibrio anguillarum.

Symptoms

1. Formation of extensive dark blood coloured areas on skin.

2. Eye may be affected, first sign is corneal opacity which may lead to ulceration in orbital contents.

3. Internally: The main feature is the enlargement of kidney and other visceral organs.

Control

In oxytetracycline and sulphonamides or nitro furans are the commonly used drugs.

II. Viral Disease

1. Carp pox

2. Infection pancreatic necrosis (IPN)

3. Lymphocytosis

4. Channel cat fish virus (CCV)

Carp Pox

Causative Agent
Herpes species.

Symptoms
1. The main symptoms are the occurrence of small milk white spots or lesions on the most parts of the body. These white lesions will gradually grow in size and cover most of the skin part. These lesions are 1-2 mm above the surface of the skin.
2. Under heavy infection normal growth of fishes inhibited and showed skeleton weakness, owing to incomplete ossification of the bones.

Control
Treatment is lacking. Avoidance is the only control.

Infections Pancreatic Necrosis (IPN)

Symptoms
1. Affected fish is darker in colour and often shows a characteristic spiral swinging behaviour associated with nervousness.
2. Exophthalmoses (swelling of eye) and abdominal distension.
3. Anterior abdominal mass containing the pancreatic tissue show the hemorrhage.
4. Liver and spleen are often swollen and ingestion.

Control
There are no means of treating infested fish to eliminate virus. Avoidance is the only effective control measure by destroying the infected fish.

Lymphocytosis

Symptoms
1. Affected fish develop small pearl like tumefaction either singly or in groups on the skin of the body.
2. Lesions are found on fins and tail and less commonly on gill filaments, pharynx, intestinal wall mesentries, liver, spleen and ovary.
3. These lesions increase in size and also become fibrous and latter on skin may develop sand paper consistency or out growth may be formed on the fins.
4. Internally entire peritoneum and pericardium may be covered with white lesions.

Control
Infested fish cannot be treated, better to destroy it.

Channel Catfish Virus (CCV)

This channel cat fish virus is a herpes virus group. The disease occurs in fry and fingerlings during summer months.

Symptoms

1. Loss of equilibrium, spiral swimming movement and hanging vertically in the water.
2. Blood oozing from the gills, skin and internal organs.
3. Lesions appear to begin from the posterior part of kidney.
4. Necrotic lesion develop in liver, spleen and digestive tract.

Control

Treatment is not possible. Hence, control by avoidance is the only possible measure as present.

III. Fungus Disease

The following are the common fungal disease in fishes:

1. Saprolegniales
2. Gill rot or branchiomycoses.

Saprolegeniales

Causative Agent

Causative agent is *Saprolegenia parasitica*. This species attack to the fry, fingerlings and adults of major carps. The infection is secondary and it infests on injured parts.

Symptoms

1. Fish becomes weak and lethargic and gradually dies after ulceration on skin and gill.
2. Hemorrhages occur, blindness, congestion at the base of pectoral and anal fin.
3. Tuft of hair like structure or out growth may occur in the affected region.

Control

Dip treatment in 3 per cent NaCl or 1 : 2000 $CuSO_4$ Solution or 1 : 1000 $KMnO_4$ for 5-10 minutes for 3-7 days.

Gill rot or Branchiomycoses

Occurred in Summer and also due to excessive organic manure.

Causative Agent

Branchiomyces sanguineus or *Branchiomyces demigraus*.

Symptoms

Obstruct the Vein of the gills, red flaking on gills, followed by grayish white, finally filament disappear.

Treatment
1. Stop feeding, no managing pond filled with freshwater addition of quick lime @ 50-100 kg/ha.
2. Dip treatment 3-5 per cent NaCl solution for 5-10 minutes for 7 days.
3. In 5 ppm $KMnO_4$ solution for 10 days.

IV. Protozoan Disease

Various protozoans infect the fish. Most of them are ectoparasistes but few species are endoparasistes. Protozoan attacks on gills and internal organs and muscles. The following are some of the important protozoan's disease.

Chthyophthiriusiasis (White Spot Disease)

Causative Agent
Ichthyopthirius multifillis

On the blood capillaries and smoothness by distressing the times resulting in the small bladder like swelling is developed. Due to their repaid respect of multiplication they damage the whole skin which actually results in death of fish.

Symptoms
1 Formation of cysts on the body surface cause simple hyperlapsia of the epidermal cells around the site of infection causing formation of pustules.
2. Heavy infected fish become a dirty grey in colour prior to death, which is due to failure of osmoregulatory processes.
3. Heavy infected fish become a dirty grey in colour prior to death, which is due to failure of osmoregulatory function.

Control
1. 10-15 minutes dip in 1,250 formalin solution.
2. Dip in 3.5 per cent Sodium chloride solution for 10-15 minutes.
3. Infected ponds should be disinfected with quick lime application.

Costiasis

This disease is characterized by formation of slimy secretion on the skin.

Causative Agent
Costia necatrix.

Symptoms
1. Sliminess can be observed easily like thin grey bag and colour becomes pale as the slime covers over the skin.
2. Fins are usually folded.
3. Heavy infected fish rub themselves against the substrata, losing of scales and opens the way force secondary infection.

Control
1. Bath treatment in 3 per cent NaCl for 10 minutes.
2. Dip in 1:2500 formalin solutions.

V. Disease of Worms

Under this group normally annelids (leech) platyhelminthes and nemathelminthes are found.

Annelids (Leech)

Leeches are the common annelids which parasites the fish, they are echo parasites and blood suckers. They attack all fish in pond water. Actiological agent is *Piscicola geometrica*. These have cylindrical body with disc like suckers present at anterior and posterior ends, which helps in sucking the blood.

Symptoms
1. Affected fish may be covered with parasites.
2. Fish become anemic and under heavy infection dies.

Control
1. Bath in Lysol solution 1:5000 for 5-10 seconds.
2. Quicklime 2 mg/litre for 5 minutes.
3. $KMnO_4$ disinfection or glacial acetic acid.

VI. Helminthes Parasites

Helminthes parasites can be broadly categorized into four groups namely:
1. Monogenetic trematodes
2. Digentic trematodes
3. Cestodes, and
4. Nematodes.

Monogenetic Trematodes

Owing to their simple life cycle where an intermediate host in involved, they can multiply or reproduce infection to fishes. The important general monogenetic trematodes are:
(i) Gyrodactylus, and
(ii) Pactylogyrus.

Gyrodactylus

Gyrodactylus is a serious infection in fishes. They are strictly host specific. They affect the skin, gills and fins.

Symptoms
   1. Fading of colour of skin, drooping of scales, occurrence of excessive mucous on caudal peduncle and fins.
   2. Skin ulcers and damaged gills.

Treatment
   1. Salt bath in 25 g/l of water for 10 minutes to adult fish and for fry 15 mg/l for 20 minutes.
   2. Formalin bath 0.25 to 0.50 ml/l of water for 30 minutes.

Diseases by Digenetic Trematodes

Diplostomasis or Black spot disease: Causative agent is *neodiplostomun* species.

Symptoms
Development of black spot on several parts of body, sometimes on eye and fins, also.

Control
   1. Dip treatment in 3 ppm picric acid for one hour.
   2. Removal of snails present in pond.
   3. Cestodes disease (Tape worms).

Ligulosis

Causative Agent
   Ligula intestinalis

Symptoms
   1. Swelling of belly, extent of swelling depends on the numbers of parasites.
   2. Reproductive organs may destroy, so that the fish becomes infertile.

Control
There is no exact treatment but it can be controlled by destroying damaged fish.

B. Non Parasitic Diseases

The main non-parasitic affections diseases are as follows:

1. Malnutrition

For fish nutrition, a number of vitamins are required in its diet; only the requirements may vary in different species of fish and depend on environmental conditions. Avitaminosis may also play a role in infections diseases by lowering the existence of various bacterial infections.

2. Intoxicant

Many adverse conditions of a chemical nature can affect the health of fish. Extreme acidity or alkalinity of water may exert a direct and specific action. Many industrial west disposals contain fish poison, such as phenol and chlorine compounds, Arsenic and heavy metal compounds, which are real threat to fish population in industrial areas. Poisonous substances, however, may also be produced by metabolic processes. Among them are NH_3, saturated hydrocarbon and Nitrides all of which result from microbial breakdown of proteins in an aerobic condition.

3. Temperature

Higher temperature beyond limit would cause exhaustion for the fish, solubility of oxygen in water is also dependent on temperature. Higher the temperature lesser is the solubility of gases.

4. Light

The growth and maturation rates of fishes are also controlled by light by photoperiod. Light has a primary role in food production. Excess of light stops photosynthetic action and may cause sunburn to the fish.

5. Dissolved Gases

Super saturation of oxygen cause the "Gas embolism" and depletion of oxygen in water results "Asphyxia" of the fishes. Minimum requirement for the good growth of fish is 5 mg/l.

6. pH

Best growth of fish is expected in water having a pH range in between 7.0–8.6. In high pH toxic, NH_3 is formed which is toxic for fishes. Free NH_3 in water should not exceed 0.02 ppm for healthy growth of fish. In low pH or Acid water, CO_2 attains toxic limit and cause fish mortality.

7. Nutrition

Nutritional diseases of fishes are generally encountered in intensive fish culture where wrong formulation or deficiency of vitamins are the other factors for disease manifestations, leads in improper growth, anemia etc.

Chapter 20

Techniques of Seed Production of Air-Breathing Fishes and their Culture

The scope of exploitation of the fishery potential of the extensive swamps and other dedicated water masses in our country is estimated to be over 1.07 million ha in areas.These vast areas are not amenable for carp culture and their reclamation for the purpose will be more expensive and the immediate returns may not compensate the investment.

It has been possible to develop techniques to culture the commercially important air-breathing fishes such as *Murrels, Clarius batrachus, Hetropneustes fossilis* and *Anabas testidineus* etc. in our country.

However, there is no trade of seed of air breathing fishes in the country. After the techniques of Air-breathing fish culture has been developed, it became necessary to organize the collection and transportation of stockable materials of air-breathing fishes for controlled culture and production.

The important air-breathing fishes with commercial values are Magur (*Clarius batrachus*), Singhi (*Hetropneustes fossilis*), Kawai (*Anabas testidineus*) and Murrels (*Channa* Species)

Magur (*Clarius batrachus*)

Magur breads during rainy seasons. During this period, it migrates from its habitual ponds to nearby inundated low-lying areas for spawning. For, laying eggs it generally constructs nests in the form of horizontal holes in the muddy embankments about 30-50 cm below the water surface collection of fry from these holes is a very simple job but the real difficulties lie in detecting the nest in the natural habitat of the fish. The collection of fry and fingerlings of Magur is quite impossible by netting operation. The only way to collect them is by getting the site dewatered wherever their presence is suspended.

Paddy fields being one of the most preferred breeding ground of Magur and better scope for availability of seeds the usual technique of collection in such a place is just to fix up a trap at the outlet of the paddy fields and allow the flow of water to stream through it. The young one of the fish coming along with the current is filtered in the trap.

Seed Production through Induced Breeding

It was noted that the low survival of developing eggs and very poor yield of hatching are the major problem of Magur breeding in captivity. It was observed that the stick eggs of the fish get fatally injured by the movement of spawners in the limited breeding space. Its breeding behaviour is very lengthy process and continues upto 10-12 hours. During this period repeated mating few number of eggs are extruded at each mating.

Paddy field remains the most preferred breeding grounds of fish in nature. This is why the efforts are made to breed Magur in prepared paddy field.

The general methodology of induced breeding is the same as it is in the case of Indian major carps.

The dose being 10-12 mg of carp gland 100 g body weight of the recipient. The only change is the use of paddy field in the place of breeding containers. The construction of paddy field should be as such:

1. Dykes should be strong.
2. Source of water to be arranged for maintaining the water level through out the course of operation.
3. Appropriate fencing has to be provided to prevent the migration of spawners, and
4. Such strains of paddy should be grown to tolerate deeper water.

The result of experiment regarding to the rearing of fry to a stockable size is still irregular and problems of survival are yet to be very less. The survival of larvae seems to be connected mainly with the availability of right type of feed after their yolk sac is absorbed. The minute plankton served as the best food for fry on an average 1500 to 2000 fry could be reared in a small area measuring 1 × 3 × 0.6 m without mortality. The rearing of 20000 fry of mangur required a basket of 1.2 × 1.5 × 0.9 m and fed them with three litres of zooplankton every morning and one kg of fish flesh and 250 gm of peanut cake in the afternoon. After two weeks, the fry had grown from 1.5 c.m to 4.3 c.m in total length. In the rearing experiment conducted that the nylon haps has so far given the most satisfactory result.

In this method the nylon hapas with appropriate mesh size are fixed in a pond and the fry is stocked in suitable density. The obvious advantages of this technique are:

1. It provides abundant zooplankton to the growing fry.
2. Keeps them safe from predator.
3. Provides suitable water column for convenience in their newly acquired aerial respiration, and
4. Permits manipulation of their population density while rearing.

Singhi (*Heteropneustes fossils*)

Singhi breeds in confined water during monsoon months. Unlike Magur, Singhi does not migrate for breeding from the natural condition. The fry and fingerlings are caught with the help of earthen pitchers. In this method earthen pitchers are placed in the natural habitats of the fish. In due course of time colonies of Singhi start utilizing this installed pitchers as their nests. The pitchers are then taken out from the place where this were kept after a interval of time. One pitcher may give yield of more than 100 Singhi.

Seed Production through Induced Breeding

Singhi preferably breeds in captivity to avoid the risk of Cyclops attack upon its young ones. In this respect larvae of singhi stands much weaker to those of Magur. The dose of pituitary gland for breeding is the same as magur as 10-20 mg/100 gm of body weight. Breeding takes place 10-12 hours after the injection, normally a female weighing 100 gm producer about 8000 eggs. The incubation period is 18-20 hours at 26-29°C. Yolk absorption takes place on the 4th day of development. Rearing of hatchlings poses no problem till the yolk sac absorption stage for mass scale rearing. Early post larvae are reared in nylon hapa of 50 mesh/linear cm fixed in a pond. The small mesh helps in filtering out the Cyclops injurious insects effectively and at the same time allows the growth of minute harmless planktonic organisms inside the hapa upon which the stocked larvae can be fed. It has already been possible to achieve 50-75 per cent survival of post larvae to a size of about 20 mm within 15-20 days of rearing through this process. Further, rearing is done in hapas with bigger mesh the zooplankton population in hapas is regularly watched and the dearth, if noticed is met from outside. Within 15-20 days fry are able to achieve the stock able size of 50 mm.

For rearing purpose cemented cisterns also give good result if prepared properly. For this a layer of mud about 5 cm is provided at the bottom and aquatic weeds are planted after maundering the water with cattle dung at 10000 kg/ha. The water level of cisterns maintained around 45-50 cm. Fry are stocked at the rate of 1500 number/m within a month young ones become suitable for stocking.

Koi/Kawai (*Anabas testudineus*)

Natural Source of Seeds

The natural spawning is continued in the rearing season. During this period, it lives in habitual ponds and migrate to the nearby inundated areas for spawning. During rainstorm it is generally found that *Anabas* prowls about on land aided by the stiffy edge of its opercula. For laying it does not construct nest. It scatters free floating eggs. The best period of seed collection is immediately after rain when paddy field and pools still remain inundated. Seeds are collected by using cast net having small mesh size.

Seed production through induced breeding of kawai can be bred easily in aquariums in collected rain water or by simply controlling the temperature of the ambient water.

For breeding through hypophysation a low dose of 15-20 mg of carp glad irrespective of the size of the fish has been generalized as a routine technique. Breeding starts after 6-14 hours of injection. On an average a female weighing about 60 g produce about 20000 eggs. Egg hatch 10-20 hours after spawning. Rearing of larvae is done taking into consideration its susceptibility to attack of Cyclops interbreed cannibalism and space requirement.

By adopting appropriate measures it has however been possible to rear the induced bred fry of kawai to stockable size with a survival of 50-60 per cent. The salient points of rearing techniques are

1. On an average 1.5 litres of space has to be provided to each fry.
2. Abundant supply of plankton has to be maintained.
3. The "shoot" fry has to be removed from time to time.

Murrels (*Channa* species)

The cultivable species of murrels are *Channa striatus*, *Channa marulius* and *Channa punctatus*. All the three species are pond breeders and are known for parental care which assumes the form of building a nest for the protection of their brood. The spawn or broods of Murrels occur in great abundance in shallow rain filled puddles, ditches as well as in tanks, ponds reservoirs, rivers etc. infested with weeds. Collection of murrel seeds should always be timed with their availability in early fry stage because they grow bigger. They loose their shoaling tendency which makes the collection difficult at that stage.

Seed Production through Induced Breeding

The murrel larvae are difficult to rear in laboratory it is suggested that the rearing of young one through this stages may be left to nature and the fry ¾–1" long collected from natural habitats.

The breeding of air-breathing fishes has now been possible to breed by hypophysation methods successfully. All the three species of Murrels are bred in large scale. The range of dose of carp pituitary glands for male is form 15-30 mg/kg body weight and for female 80-120 mg/kg body weight of the fish.

Culture of Air-Breathing Fishes

Pond Culture

A large number of shallow swampy ponds are available in the country. Such ponds are some times used for culturing makhana (*Euryale ferox*) as a cash crop. These ponds can also be used for the culture of air-breathing fishes *viz.*, Singhi (*H. fossilis*) Mangur (*C. batrachus*) and Koi/Kawai (*A. testudineus*). In an experiment a pond, measuring 0.04 ha, was stocked at the rate of one lakh fingerlings/ha of water, stocking ratio being Singhi 7.3 Magur 2.1 Koi 0.3. When the average weight of the specimens where 9 gm, 10 gm, and 12 gm respectively. The pond was neither fertilized nor the fishes were fed. 2500 kg of fish could be harvested per ha of water after 7 month rearing of fish. In a similar experiment when a 0.03 ha pond was stocked with Singhi and Kawai/Koi @ 10000 each species/ha of water a net production of 524 Kg fish could be harvested, when the gross weight of fish restricted to 1200 kg/ha.

Air-breathing fishes are either carnivorous or omnivorous in habit. They readily accepted supplementary feed which generally a mixture of rice bran, ground nut oil cake with dried silk-worm pupa, dried marine trash fish. Slaughter of house waste can above be used as feed of Singhi and Channa species. Cost of this material when collected from the source is quit low.

In towns and village, small water bodies are plentiful and may be utilized for air breathing fish cultural. The pond may be stocked either with murrel or Magur, or a mixture of Magur, Singhi and Koi. The stocking density of murrel is 20,000 fingerling/ha, Stockable size of these fishes may be 6-8 cm. in mixed cultural, the density may be restricted to 50000 fingerlings/ha. For semi intensive cultural, the stock is to be fed twice daily before dusk. The quantity of fed per day is of the tune of the 3 per cent of the body weight of the stockable fish. To determine the quantity of feed required a few fishes are to be caught and be weighted fort-nightly when Murrels are fattened for 8 months and Magur for 6 months an yields of 5000 kg/ha of water can easily be expected.

The Economics of Magur (*C. batrachus*) in Six Month Culture

(1) Lease value of derelict water are for 6 months	Rs. 10000.00
(2) Feed (dried marine trash fish @ Re 10/- per kg)	Rs. 10000.00
(3) Rice bran @ Re 2/-per kg	Rs 2000.00
(4) Cost of Fingerlings (40,000 Nos.@Rs1/-)	Rs. 40000.00
(5) Labour for feeding and Guard etc	Rs. 24000.00
(6) Bank interest @10 per cent on the capital of Rs.86000/-	Rs. 8600.00
Total in	**Rs. 94600.00**

Income

1) Yield 5000Kg/ha @ Rs. 100/Kg = Rs.500000.00

Net Profit is Rs. 500000.00 − Rs. 94600.00 = Rs. 405400.00

Culture of Magur in Cisterns

Culture of Magur in backyard or cisterns is proposed as a small scale endeavor. In a cistern, measuring 3m × 1m × 0.75m at least an average weight 10 gm of fish can be reared and fattened. The stock should be fed a supplementary diet of a mixture of 1 part of groundnut oil cake with 4 parts fried and ground plash marine fishes every morning and afternoon. One gram of yeast should be mixed with 1 kg feed to enhance growth of the stocked fish. The fed caught to weight 1.5 per cent of the total weight of the fish when the stock is reared for 6 months a table sized fish weight ranges between 75 to 125 gm can be harvested. The water of the half-filled cistern is to be changed daily in the morning before the fed is given. This profitable enterprise brings pleasure for the family as well.

Culture of Air-Breathing Fishes in Cages

Air breathing fishes can also be cultured in cages to utilize vast swampy weed infested waters, which pose harvesting problems. A cage of 1.5 × 0.75 × 0.75 m made of split bamboo mats can held 200 fingerlings of any species. Half the top of the cage is kept covered with a piece of net instead of bamboo mats to release feed in the cage. The stock is fed twice daily. The ration is not more than 3 per cent of the total weight of the fish stocked.

The feeds is a mixture of rice barn with slaughters house waste (1:1) are given to Singhi and Koi/Kawai. Rice bran mixed with dried silk worm pupae (1:1) are given to above mentioned species. Magur feeds on dried ground marine fresh fishes and groundnut (4:1). The net productions of cage fish will varie species wise, Singhi and Magur 3 kg/cage in 4 months and 1.75 kg/cage in 4 month. The gross production in this crops per year of different species in Singhi is 16.5 kgs, Magur 1.65 kgs and Koi 8.25 kg. Initially the economics of "Cage Culture" appears unfavourable as the cost of each cages bring Rs 100/- but the same can be used when three crops are raised in each cage. A cage can be used for two years making minor repair after every crop is harvested. It is easy for a person to manage or make 10 cages. The economics of each culture may be summed up as follows:

1. Cost of materials for the cage — Rs. 1000.00
2. Wear and tear of the cage — Rs. 500.00
3. Seed (self collected) — Rs 500.00
4. Feed (400 Kg for three crops/year) (@ Rs.10/- per Kg) — Rs. 4000.00
5. Bank interest @10/- per year — Rs. 600.00
6. Gross Income from 160 kg cage fish Produced @16 kg/Cage/Year; rate of fish sold being rupees 100/kg. — Rs. 16000.00
7. Net income (16000-6600) in a year from ten cages — Rs. 9400.00

Chapter 21

Composite Fish Culture or Polyculture or Integrated Fish Farming with Livestocks

In order to obtain high production of fish/ha of water body, fast growing compatible species of fish of different feeding habitats or different weight classes of the same species are stocked in the same pond so that all its niches are occupied by the fishes. This system of pond management is called mixed fish farming or composite culture or polyculture. The basic principle of composite fish culture is that when the compatible fishes of different feeding habits are stocked together they secure for manner, all life requirements available in the pond for fish production without harming each other. In fact, there does not exist any serious competition between different species but each species may have a beneficial influence or growth and production of the other. For example, Grass carp depends on aquatic vegetation, converts or excreta fertilizes the pond which benefits all the species.

The fast growing species of fish of different feeding habits are stocked together in the same pond, so that all the ecological niches could be utilized by the fishes in order to achieve high production. A maximum of 10,000/ha is achieved by this technology. A combination of live species of Rohu, Catla, Mrigal, Silver carp, Common carp can be done for a clean pond. Grass Carp is advised to be included in weeded ponds only. The stocking density is 5000 nos/ha. Comparing the ratio, the silver carps-2, Catla-1, Rohu-3, Grass Carp-1, Mrigal 1.5 and common carp 1.5.

The composite fish culture is viable technology for rural ponds of different sizes. Which require certain managements practices for maximizing the production such as eradication of weeds, removal of predatory fishes, use of lime, use of organic and inorganic fertilizers, use of artificial feeds, like rice bran and oil cake, periodical replacement of water etc.

The intensive composite fish farming of above mentioned management practices followed with a stocking density of 7,500 per ha. A production of 9000 to 10000 kg/ha could be achieved successfully.

Integrated Fish Farming

In the integrated fish farming, carp farming system, farm commodities are produced in a complementary manner. The excreta of animals are used as organic manure for fertilization and agricultural fish pond. In fact in integrated farming nothing is wasted. The pond is used for fish culture whereas the whole embankment or associated land can be used for vegetable and horticultural plants. This system of farming yields good dividends to the farmers. Following are the combination of integrated farming systems:

1.   Fish-cum-Duck Culture

Ducks are ideal for integration with fish culture. They draw a large share of their nutrition from fish ponds by eating aquatic weeds and insects which are undesirable elements for fish culture. Their dropping are rich in pond nutrients which lead to a good growth of fish food organisms. A production of 3700 kg/ha could be achieved by rearing 200 ducks. On the other hand which yields 15800 eggs and 500 kg of duck meat. The cost is primarily on ducklings, duck feed and labour. An investment of Rs. 60000/- per year may give a net income of Rs 45000/- by scale of products including fish when stocked with 5000 Nos/ha both of fingerlings in ratio Catla 1.0, Rohu 1.8, Mrigala 2.8, silver carp 1.5, Gross Carp 1.0 and common carp 1.9.

2.   Fish-cum-Pig Culture

Raising Poultry and Fattening Pigs in combination with Fish farming is becoming its popular method adopted in China, Taiwan, etc. In this method Pigsties are being constructed at the marginal bands of fish ponds. It is estimated that 50-70 numbers of pig with or without fermentation can supply the adequate amount of fertilizer per 1 ha pond. The production achieved is 8000 kg of fish in a year long culture along with 8000 kg of pig meat. The expenditure incurred is approximately 180000/- which give a net profit of about Rs. 300000/- in a year. The pond is stocked with 7000–8000 fingerlings. The cost of production is approximately to be Rs. 10/- per kg of fish.

3.   Cage Culture

Cage culture has been practised in Combodia rather traditionally. It has been adopted in many other countries only recently. The technique has been proved very successful, particularly for some cyprinoids, silurides, clupeds and others. Japan has come out as one of leading countries with very good results. Successful rearing of yellow tails–a pelagic clupeid–in Japan has open new grounds of exploitation in marine fish culture. The utility and importance of the technique of cage culture on commercial scale is atone clear when yields are compared. Cage culture has opened new ways of easy exploitation of natural waters, both freshwater and salt water, which are more suitable for fish raising than the waters of artificial ponds or the like. In Cage culture though the fish remain in confinement, the water available to them, and with most of the requirements is unlimited. In running nature of water which is the key to greater production. The running water keeps a constant supply of food, a regular removal of wastes and quick replenishment of what is used up *i.e.*, oxygen, minerals etc. by the caged crowded fish. Management and investment of capital is minimum in cage culture. The greater yield obtained from the cage culture is thus accounted for the following characteristics:

(a) High densities of stocking of fish.

(b) Volume available to fish being much larger than the actual area or volume enclosed in the cage.

(c) Relatively simple and cheap technique applicable to both fresh and salt water.

Cage culture is being practised at present in a number of countries and for a variety of fishes. Cage culture has proved a boon for reclaiming extensive, waste waters such as swampy water spread area, derelict waters, stagnant water, waste water etc. is about 0.6 million ha in India, where air-breathing fishes such as *Clarias, Heteropneustes, Channa* sps, *Anabus, Cyprinus carpio* (Common carp), *Tilapia* etc. are responding well to cage culture practices. Cage culture are made of split-bamboo mat supported on farmers measuring 2m × 1m × 1m. It may be recalled that harvesting of fishes raised in open swamps is an impossible task due to mud and weeds. The cage permit easy harvesting of reared fishes. Cage culture of catfishes like *Mystus sps.*, is also being explored in rivers. Oil Sardines are reared in cage in marine coastal waters.

4. Paddy-cum-Fish Culture

Paddy fields are flooded from periods varying from 3 to 8 months in a year and since some growth of fish can be achieved in this period, it appears desirable to use paddy field for fish culture also. The culture of fish in paddy fields can be divided into 3 groups:

1. Along with paddy during the period of cultivation.

2. Secondary crop after paddy has been harvested.

3. As a continuous culture that is transforming the fish in specially prepared ditches channels during the paddy harvest or when the field are drained.

When, there is sufficient irrigation facilities or a continuous supply of water can be maintained in the field can be raised as the only secondary crop. Fish may be introduced from elsewhere or wild fish brought in with the water that is let into the field the period of grow is approximately six month and fishes are harvested from October onwards ears is taken that the fish do not escape from paddy fields till the complete harvesting is over by December end. According to an estimation of fish from such field is fish from such fields is approximately 900 kg/ha.

In Kerala, approximately 1100 acres of paddy field are used for prawn culture with bankments of 5' to 6' and sluice prawns are let in along with high tide. According

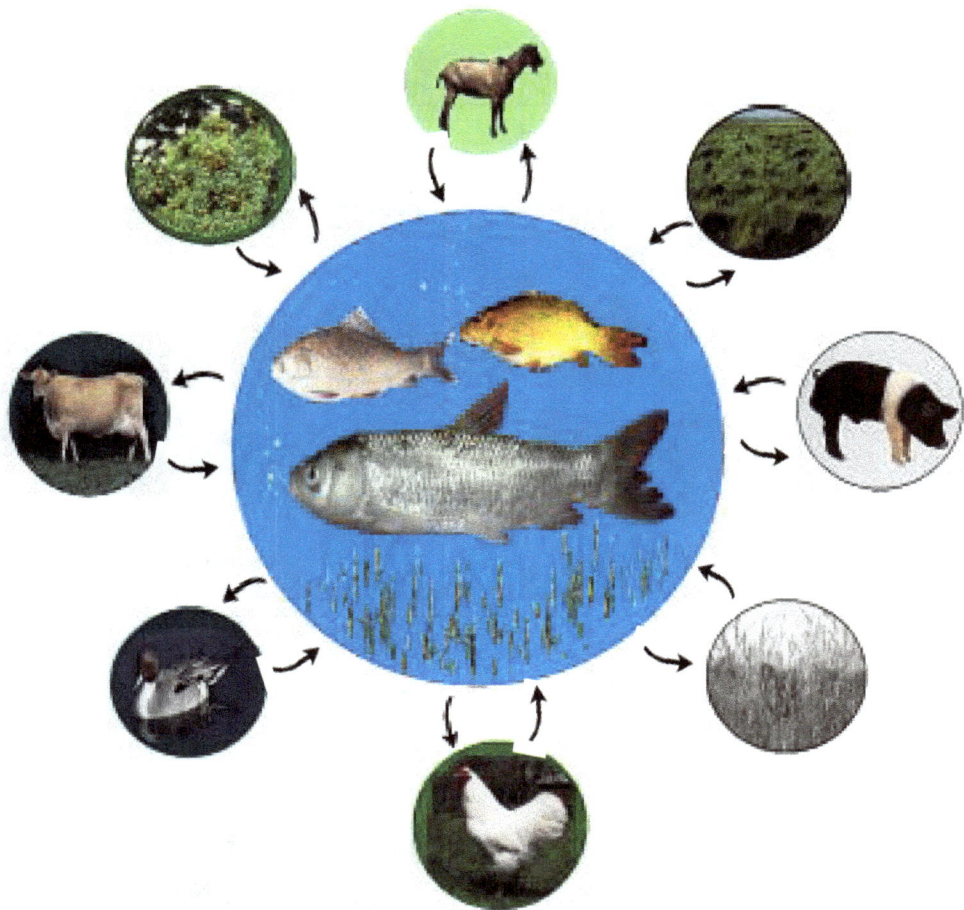

Plate 15: A Model of Polyculture

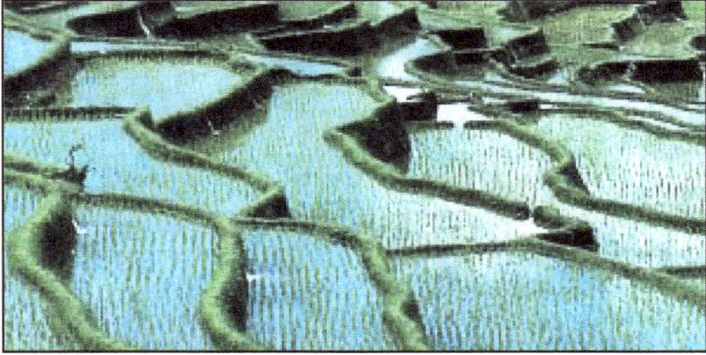

Plate 16
Showing Paddy-cum-Fish Culture at Ziro Valley in Arunachal Pradesh

to an estimate about 2000 to 5000 tones of prawns are caught from these fields annually with an average yield of 360 to 680 kg/ha.

In Andhra Pradesh million of carppy enter in the irrigational pady fields during June to July from the rivers and they grow to 4-6 cm in fields till September-October. It is a wild culture and after paddy is harvested the fields are drained out fished. The depth of fisheries has started collecting those seeds and stocking in deeper waters.

In Tamil Nadu also Paddy fields in certain areas of state gets fresh seeds like Andhra Pradesh from river and harvested after 6 months.

Water and Soil Condition

Most successful fish culture is obtained in the area where controlled irrigation is practiced where fields depend upon rain water. A depth of 30 cm of water is maintained. In shallow water temperature may raise unduly so deeper ponds or cannels are needed for the fish to survive. In blackish water of paddy fields mullets, *Tilapia* and prawn can be mixed successfully. Paddy gives good crops of fish due to its productivity.

The nitrogen content of soil in paddy fields used for carp culture and fertilized in the usual way; is also the beginning of carp rearing then it is at the end of the period. It appears at the early stage of the paddy cultivation due to the deposition of excreta of fish nitrogen increases in the fields and this does not adversely effect the growth of the paddy.

Biota

The planktonic biotas of the paddy fields are almost similar to those of the shallow ponds. Filamentous algae and weeds also grow in paddy fields and some of this vegetation are used by fish as food and kept under controlled by them. Various insects also harbors which are injurious to paddy such as mosquito and its larvae are controlled by the fish as their food.

Preparation of the Paddy Fields

1. There should be adequate water in the fields.
2. Bundh should be strong.
3. 95 cm height of bandh is essential for *Tilapias*. It should be 60 cm.
4. Old bundh should be repaired so that frog etc may not take shelter.
5. A narrow channel in the paddy field along the bundh is preferred.
6. It will be advantageous to dig a pit connected to the channel which gives shelter to fish from sun heat and it is easier to catch them there.
7. Usual width of channel is 50 cm and depth 20 cm and the pit 1m × 1m at the junction of channel with 60 cm depth.
8. Suitable outlets and inlets are also essential with adequate control device.
9. Paddy stubble should be allowed to decay to give raise to planktons.
10. Organic manures are preferred and it should be increased by 50–100 per cent when it is utilized for fish culture.

Stocking

The species suitable for paddy cum fish culture:

1. Thrive in very shallow water.
2. With stand fairly high turbidity of water.
3. Tolerance relatively high temperature and low oxygen content.
4. Grows to a remarkable size in short period.

Common carp is the most suitable for this purpose which have the ability to adjust with the above mentioned factors. *Tilapia* and *Puntins sarana* are also known to be useful. Murrels and cat fishes stocking data largely depends upon:

1. Duration of culture.
2. Productivity of water, and
3. The size of the fish induced.

The general practice to stock at the rate of 2000 per ha or 2500/ha when additional food is supplied yearlings @ 1200–1600/ha are stocked without artificial food and 4000 with supplementary food. Somewhere only *Tilapia* is stocked with Carp @ 750-1500nos./ha.

Management

1. Precautions should be taken from predatory birds.
2. Excessive growth of filamentous algae can be fatal to fry.
3. Manuring should be at the regular intervals.
4. Artificial food such as rice barn and M.O.C should be given.
5. Bundhs should be regularly inspected and the mud from channel should be cleaned time to time.
6. Water supply should be controlled and should be properly maintained.
7. It should be practiced from highly effective insecticides.

Economics

It is a subsidiary activity and should be modified to suit paddy cultivation. The income from fish culture is an additional one and provides a cheap acceptable nutritious protein food in fresh condition. Also the rice yields increases oftenly in presence of fish by 15 per cent. The actual yield depends upon:

1. Species stocked.
2. Pattern of culture.
3. Fertility of soil and water.
4. Food produced and its quality.

Problems of Integrated Fish Farming

Integrated fish farming undoubtedly generates extra income but on the other hand it needs an intensified effective management, especially one must have sufficient knowledge of animal husbandry for the culture of Ducks, Poultry, Pig etc. One has to be extra careful about changing of water quality especially dissolved oxygen content and pH. Due to constant recycling of organic waste, as animal excreta is a potential source of various parasitic diseases and need gradual check up of the stock. In fish cum duck culture, ducks are likely to prey up on the similar fishes. To avoid these fingerlings of over 4" size fry should be preferred to stock in culture tanks.

Chapter 22

High Tech Aquaculture Systems

With the fast growing population of India, it is estimated that to provide adequate animal protein, to the Indian population, about 30 million tons of fishes shall be required by 2030 A.D. and out of this inland fish culture should contribute 12 million tons. The present production from inland resources is over 6.4 million metric tons. So, nearly 5 times more fish production has to be achieved in the next 21 years. The target is very high but it can be achieved if high technologies are developed and adopted in the field of aquaculture. Recently in some fields of aquaculture, *viz.* fish-seed production, fish culture, polyculture, prawn culture and culture of live feed organisms, advanced technology etc. are developed and being adopted with success.

The present production of fish seed also need nearly six-fold increases to meet the growing demand of fish seed. As per the estimation, the fish seed demand to stock in the available inland water. 80 per cent we are only provided 80 per cent we are only provided. 40-50 per cent survival is obtained even when the spawn is stocked at very moderate rate of 0.5 to 0.6 million/ha, is improved to achieve better survival rate with higher stocking density, the above stated target cannot be achieved. Fish seed being the primary requisite of fish culture should be given top priority to increase its production so that all the available water area can be stocked with adequate fish seed for integrated fish farming. Intensive rearing systems of carp fry developed recently and it should be adopted to boost the fish fry production from a unit area in limited time.

Due to sustained efforts by the Indian fisheries scientists, the fish production from ponds has increased from nearly 1 ton to 10 tons/ha, with the development and adoption of advanced cultural techniques like composite fish culture and mixed culture. But this is not the end. But it is high time to further advance the technique to increase the tank fish production. Running water sources can also be tapped for fish culture by developing and improving the cage culture or pen culture techniques.

Reservoirs, fishery resources are in plenty in our country but it is really a matter of concern that despite massive efforts during the last five decades, the reservoir fish production could not be increased to the desired label at least 40 kg/ha, whereas, other developed countries like USA, and Russia have achieved long back. So this is the most potential field where the hi-technologies are to be developed and applied for optimal utilization of the resources.

Advances in Fish Seed Production Technology

Fish Breeding

Development to induce breeding techniques has subsequently increased the production of quality fish seed of Indian and Exotic carps by hypophysation for last five decades in India. This technique is improved in itself simultaneously. Now there is an unrest desire to find out the real substitute of fish pituitary gland due to the increase in demand and limited supply. Several drugs were tried in the past by various workers and institution in India and abroad but with a limited success. Recently the H.C.G. and homeopathic medicine are successfully used for inducing spawning in exotic carps and in combination (40–50 per cent) with pituitary glands with Indian carps by the Indian scientists. So this drug can be effectively used for gonadal development and spawning of carps. The Indian Fisheries Institutions are successfully demonstrated the growth promotion nature of H.C.G if the fish food is fortified with this drug. Ovarian drug is successfully tried for inducing spurning in fishes in India and abroad.

Hatching Techniques

Second important step of seed production operation after successful spawning is the hatching of eggs. Being the most tender life stage, it is noticed that with use of conventional double hapa the percentage recovery from fertilized eggs to spawn is very low, sometimes less than 40 per cent. Realizing the importance of hatching operation, several models like glass jar hatchery, Bucket hatchery and others were evolved and tried in the past. But none of these was ever tried on commercial scale. However, CIFE D-81 to D-85 models of hatchery, circular hatchery etc. are tried successfully on commercial scale throughout the country by the fisheries scientists and fish farmers.

Besides the above indoor hatchery systems, the circular hatchery or Chinese hatchery is worth mentioning under Hi-tech system as this model is very successful for fish breeding and hatching on commercial scales and being operated in most of the states in India. Also the World Bank hatchery systems are being propagated in different states of our country.

Rearing Techniques

The final step in seed production is the rearing of spawn to fry and fry to fingerling stage. As stated earlier under traditional system of rearing of fry in nursery ponds even at moderate stocking density of 0.5 to 0.6 million/ha the survival of spawn fry is rarely above 50 per cent and survival rate of fry to fingerlings is also not good. So due consideration is given to further improve the fry rearing technology to increase the

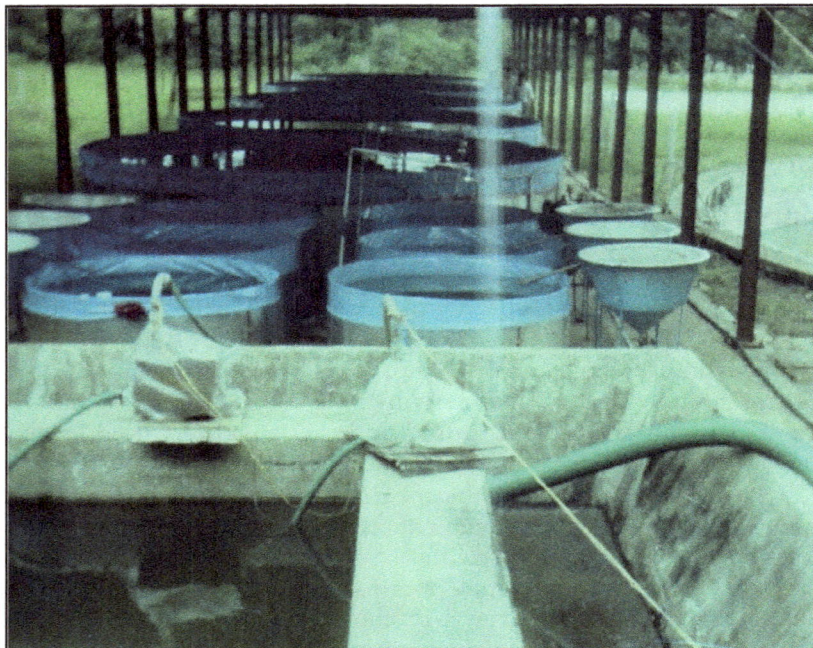

Plate 16
A View of High Tech Methods Applied for Breeding and Rearing of Fishes

survival percentage and also to increase the fry production rate from per unit area in an unit item. Presently fish seed production is 21,000 million fry and that of shrimp about 8 billion, with increasing diversification in the recent past. The total number of hatcheries up-to XI plan period is 1,070.

Feeding the fish seed with live fish feed further increases the survival and growth of seed and also keeps the ponds environment healthy. To meet the requirement of live fish feed its culture is necessary known as Zooplanktons which is very old practice but mostly utilizing the same technique on large scale. Use of artificial feed which contain essential materials shall be provided. It is proved that *Cladocerans* can also be cultured at high density varying from 10,000 to 25,000/litres.

Advances in Fish Culture

The fish production from ponds or tanks ranges 500 kg to 1000 kg/ha till the development of advanced technology like composite fish culture and polyculture techniques that increased the production up to 10 tons/ha. A new technique has to be applied as in Japan and other developing countries of South East Asia and China.

If running water culture is adopted as in Japan, where by this method, 250-300 kg/m^2 fish are produced. India can also increase its production and can achieve self-sufficiency by 2030 A.D.

Polyculture is the recently adopted method and technique where fish culture is coupled with duck farming, poultry, piggery, paddy-fish-culture, horticulture and diary etc. to boost the income of farmer and reduce the cost of inputs like manure etc. which are produced by itself by other livestocks under the intensive fish culture.

Chapter 23
Sewage Fed Fisheries

Sewage is one of the neglected resources of the world. It has been estimated that about 0.5 tons of urine and faces are produced per persons per year and another 50 tons of water is used to wash. Increasing interest has been shown at present for proper utilization of domestic wastes not only from the point of its proper disposal but also as a bare necessity to rise this rich organic manures due to acute shortage and high price of conventional fertilizers. It has the high mineral value containing the nutrients such as nitrogen, phosphorus, calcium, potassium etc.

The fertilization with domestic sewage for fish culture in freshwater ponds has been recognized in many countries. In India, there are more than 150 sewage fed fish farms covering an area of 14000 ha. Almost all of them are fed with raw or semi-treated sewage for the purpose of fertilization.

West Bengal is perhaps the only state in the country where the city sewage is being extensively utilized for culture purposes. In most of these fisheries, fingerlings between 3" to 6" in size are stocked and fish farmers have limited facilities to produce fish seed in nurseries fertilized with sewage effluent.

Sewage is fully or partly decomposed organic matter diluted in various proportions with water used for domestic purposes and carried away residences, business buildings for final disposal. The process generally adopted for the use of sewage treatment before release in fish ponds are:

1. Sedimentation
2. Dilution, and
3. Storage.

The function of sedimentation process is to remove suspended solids from sewage to the maximum possible extent. It is done by letting sewage into a tank at a high velocity of flow. Sedimentation results due to sudden drop in velocity when the sewage enters a large pond from a sewage channel. Sedimentation is carried out in two successive stages. They are:

1. Primary
2. Secondary

The primary stage is intended to settle down most of the heavy solids while the secondary stage serves two purposes:

1. Provision for additional detention period to help mix and homogenize variations in the flow and promotion of natural purification process. It has been estimated that about 33 per cent BOD is to get rid of sedimentation process which may effect 90 per cent settlement of suspended solids and about 25 per cent reduction in albuminoid ammonia.

2. Before introduction of sewage into any fishery, its dilution by freshwater should be so effective that a passive dissolved oxygen balance is maintained and the concentration of unwholesome ingredients such as CO_2, NH_3, H_2S etc. is kept below their lethal limits. The oxygen required for biochemical reaction is obtained from freshwater used for dilution and through green algae and and vegetation in the water. During the storage process the logical processes carried out by micro-organisms present on the raw sewage oxidize it.

The average chemical properties of domestic sewage in some sewage treatment ponds which is utilized for rearing of carps and tilapia are as follows:

1. pH: 6.7 to 7.3
2. Dissolved oxygen: 0.00 to 1 ppm
3. Alkalinity: 214 to 520 ppm
4. Phosphate: 2.2 to 26.5 ppm

Pond Preparation

The ponds utilized for rearing carps seed are generally dewatered during summer months. In ponds where complete dewatering is not possible, application of mahua oil cake of about 250 ppm or Eldrin of 0.10 ppm are recommended to remove carnivorous fishes from the pond.

Before the out break of monsoon in the month of April or May, fresh sewage affluent is taken into the pond upto a depth of 3' for initial fertilization. During fertilization of ponds with raw sewage, the water is heavily polluted with putrisible organic content of the sewage and cloudy appearance and disagreeable odour of the water indicates general depletion of dissolved oxygen. Masses of gaseous shidge, rearing from the pond bottom are often noticed floating near the surface of the water. The bacterial load of the pond increases very heavily during this period. The bacterial present are mostly non pathogenic and are of two types:

1. Aerobic, and
2. Anaerobic.

Aerobic bacteria require oxygen either from air or from water for oxidation, while anaerobic bacteria decompose matters in the absence of air rendering them partly liquid and partly gaseous. The sewage ponds are generally designed as aerobic and anaerobic depending upon the active conditions of the two kinds of bacteria.

During the period of fertilization for stocking carp seed most of ponds are found to be predominately aerobic during the sunshine hours as well as during the early hours in the night. During late night hours, the upper layers of the ponds may or may not be aerobic but the bottom layers are generally anaerobic. During aerobic condition, the organic matter present in sewage is oxidized into simple in offensive substances by bacteria which derive their supply of oxygen from the dissolved oxygen of the water and as a result of this, carbondioxide and ammonia are produced. Due to the release of nutrient rich organic matter into the pond, healthy bloom of algae develops. The oxidation is achieved in nature by a complete symbiosis of algae and bacteria. The algae produce the oxygen required for bacterial respiration and the bacteria, utilizing this oxygen, oxidize the organic matters in the sewage into organic forms suitable for algal utilization. Soon after fertilization of rearing ponds it has been observed that during day time the dissolved oxygen content of water reaches above the saturation level while at night the respiratory activity of the algae and oxidation of organic matter reduces the dissolved oxygen content of water resulting in anaerobic condition at the pond bottom. Therefore, the day and night variation of dissolved oxygen in pond is very great ranging from day saturation to night depletion. However, within a month, the range of variation of dissolved goes down due to depletion. However, within a month, the range of variation of dissolved oxygen and natural purification process render the pond suitable for rearing fish.

Stocking

After the ponds are stabilized, it is stocked with Indian major carp *i.e., Catla catla, Labeo rohita, Cirrhinus mrigala and* exortic carp *Cyprinus carpio*. The stocking density usually varies from 70000 and 150000/ha depending on the size of the fry. In one experimental pond about 60000 fry/ha having 40.14 per cent *Catla*, 6.23 per cent *Rohu*, 4.68 per cent *Mrigala* and 8.93 per cent common carp are stocked. The average length-weight of *Catla, Rohu, Mrigala* and Common carp at the time of stocking are 72 mm/5.9 g, 72.2 mm/hg/5.9g, 72.2 mm/4.9gm, 74.2 mm/hg/4.0g and 14.6m/2.5g respectively. The pond s are fertilized with sewage effluent at the rate of 4500000/ha. The fishes are reared for a period of 25 days and no artificial food or sewage effluent are given during the period of rearing. The average size and weight attained by *Catla, Rohu, Mrigala* and *Common carp* are 133.2 mm/30gm, 147 mm/37g, 125.7 mm/24.0 gm and 135.0 mm/50.0gm respectively. The net production achieved by this short term rearing are 145554 kg/ha. The fingerlings are then transferred in bigger stocking ponds for culturing with domestic sewage effluent.

Sewage is rarely used on rearing carps seed in nursery ponds. It was observed that fry grow into pond using sewage effluent showed a higher rate of growth than the same grown in regular nursery ponds.

Chapter 24
Prawn Culture

Indian aquaculture has been evolving from the level of subsistence activity to that of an industry. This transformation has been made possible with the development and standardization of many new productions and associated techniques of input and output subsystems. In recent years aquaculture has created great enthusiasm and interest among entrepreneurs.

Although India has vast freshwater resources they are not fully exploited except for carp culture in limited scale. Freshwater fish culture employing composite fish culture technology has become popular for use in large number of tanks and ponds in the country. To meet the raw material required by the processing units for export demand there is urgent need to expand our production base. In addition it is always stressed that there is a need to utilise our natural resources productively to ensure the much needed food security.

Scope for Freshwater Prawn Culture

Considering the high export potential, the giant freshwater prawn, Macrobrachium rosenbergii, the scampi, enjoys immense potential for culture in India. About 4 million ha of impounded freshwater bodies in the various states of India, offer great potential for freshwater prawn culture. Scampi can be cultivated for export through monoculture in existing as well as new ponds or with compatible freshwater fishes in existing ponds. It is exported to EEC countries and USA. Since the world market for scampi is expanding with attractive prices, there is great scope for scampi production and export.

Technical Parameters

The giant freshwater prawn is suitable for cultivation in tropical and subtropical climates. It is a hardy species by virtue of its ability to adapt to various types of fresh and brackish-water conditions. It accepts pelleted feed and has omnivorous feeding habit. In the natural environment, lower reaches of rivers, tidal inlets, where water is directly or indirectly connected with sea are their preferred habitat specially during spawning. The breeding takes place in low saline waters which is also needed for larval and post larval development after incubation.

Though seed may be available in natural sources to a limited extent, for large scale culture there is a need to ensure regular supply of seed. For ensuring availability of quality seed in predictable quantity freshwater prawn hatcheries should be encouraged, technology for which is already developed. Freshwater prawn hatcheries are coming up in many states of India.

Site Selection

The site selection plays an important role as the entire management aspect of the farm ultimately depends on specific conditions of the site. The aspects to be considered are topography of the area, soil type, availability of quality water etc. The area should be free from pollution and flooding. Other considerations like approach roads etc. have also to be taken into account.

Soil Quality

The ideal soil for Macrobrachium culture should be clay silt mixture or sandy loam comprising of 60 per cent sand and 40 per cent silt with good water retention capacity.

Pond Construction

Rectangular ponds are suitable mainly from the harvesting point of view. A convenient width is 30-50 m, whereas length of the pond depends on site, topography and farm layout. Normally a size of 0.5 to 1.5 ha is found suitable. The average depth of the ponds should be 0.9 m with a minimum of 0.75m and a maximum of 1.2 m. Dike and pond slope may be kept at 2:1. Bund must have a freeboard of at least 60 cm above the highest water level in the pond. Designing and layout of the farms may be done keeping in view the water intake and water outlet facilities. The drainage system should be designed carefully to prevent mixing of outlet water with incoming water.

Water Supply and Drainage

Appropriate water supply and drainage systems have to be designed keeping in view the water source and topography of the area. Tubewell and pumping system may be considered if required for water intake/exchange. Water exchange on weekly or fortnightly basis as required is desirable and provisions are to be made accordingly.

Water Quality

There should be availability of abundant and good quality water. The water should be free from any kind of pollution. The pH should be maintained at 7 to 8.5. The temperature should range from 18°C to 34°C. Dissolved oxygen content should range between 6 to 10 ppm.

Farm Management

The type of pond preparation to be adopted before stocking is based on the type of culture and its intensity and nature of the culture pond. Liming of the pond assumes great importance here than in the case of freshwater fish culture. The application of fertilizers is restricted in case pelletised feed is used. However, occasionally cow dung, single super phosphate, urea etc. can be applied on assessing the productivity.

The stocking density normally varies from 40000 to 50000 nos. of post larvae per ha depending on the type and intensity of the management practices. The culture system may be monoculture or polyculture with carps. In case of polyculture with carps the more pond depth is preferred at 4-5 feet. In case of polyculture the stocking density of prawn may vary from 15000-20000 post larvae. The carp fingerlings may be of the order of 3000-5000. Nursery may be incorporated where the post larvae obtained from hatcheries could be reared for a period of 4-5 weeks till they attain 40-50 mm or 1-3 g.

In order to get desired production, feeding, aeration, water exchange, periodic monitoring should be continued. The quality and type of feed is based on culture system. Macrobrachium with its omnivorous feeding habits can make use of a variety of feeds from common wet feed made from rice bran and oil cake to scientifically formulated pelleted feed. The rate of feeding is determined by the stage of growth of prawn, water quality, density of stock and other manuring practices. Generally the feeding rate my be 5 per cent of the body weight.

The duration of culture varies from 6 to 12 months depending on the type of culture practice. Generally in monoculture the culture period may be 6-8 months under monoculture and 8-12 months under polyculture. The average growth of prawn may range from 50 g to 200 g depending on the duration, density, water quality, feeding etc. The survival rate may range 50 per cent to 70 per cent depending on the type of management practices.

Extension Services

The borrower should have experience in prawn farming and should be conversant with production technology, trade etc. Fish Farmers Development Agencies (FFDA) have been established in almost all districts for providing necessary training. The offices of Marine Products Export Development Authority (MPEDA) in most of the coastal states also provide necessary assistance.

Marketing

There is good demand for freshwater prawn in both local and international markets; as such there may not be any problem in marketing the same. Freshwater prawns can be sold directly by the farmers either in the market or to exporters for processing before export.

Financial Outlay

Details for the financial outlay have been indicated in Table 11. It can be seen there from that the capital cost for a 1 ha unit in the first year has been estimated as

Rs. 3.045 lakh while the operational cost for one crop works out to Rs. 2.45 lakh. The items and cost indicated under the model are indicative and not exhaustive. While preparing projects for financial assistance the costs have to be assessed taking into account actual field conditions.

Table 11: Economics of Grow-Out Production of Prawn in First Year

Sl.No.	Item	Amount in Rs.
A.	Expenditure	
1.	Construction of pond	80000.00
2.	Prawn seed @ 60,000/ha @ Rs.1000/1000 nos. including transportation cost	60000.00
3.	Fertilizers and lime	3000.00
4.	Supplementary feed (3 tones@ Rs. 20/- per kg)	60,000.00
5.	Wages (One @ Rs. 2500/month for 9 months)	22500.00
6.	Electricity and fuel	4000.00
7.	Harvesting charges	6000.00
8	Miscellaneous expenditure	5000.00
	Total	**240500.00**
B.	Total Cost	
1.	Variable cost	240500.00
2.	Interest on variable cost (@ 10 per cent per annum)	24050.00
	Total	**264550.00**
C.	Gross income	
	Sale of big size prawn (@ Rs. 250/kg for 1000 kg)	250000.00
	Sale of small size prawn (@ Rs. 125/- kg for 500 kg)	62500.00
	Total	**312500.00**
D.	**Net Income (Gross income – Total cost)**	**72000.00**

Margin Money and Bank Loan

The entrepreneur is expected to bring margin money out of his own resources. The rates of margin money stipulated are 5 per cent for smaller farmer, 10 per cent for medium farmer and 15 per cent for other farmers. For corporate borrowers the margin stipulated is 25 per cent. NABARD could consider providing margin money loan assistance in deserving entrepreneurs.

Rate of Refinance

NABARD provides refinance assistance for freshwater prawn farming to commercial banks, cooperative banks and Regional Rural Banks. The rate of refinance is fixed by NABARD from time to time.

Recycling of Pond Nutrients for Plantation of Crop

The manure value of pond mud is usually considerable. A large quantity of nutrient remains locked in the sediment and not utilized in the rapid nutrient cycling of the water column in fish ponds. Thus, pond mud may be removed or scraped and can be profitably utilized for horticulture or other agricultural crops. Removing the soft sediment by some mechanical, manual or biological means also improve the productivity of the pond. Further, irrigating terrestrial crops with pond water is much better than using running water, since impounded water is more fertile and adds nutrients for the growth of plants.

Chapter 25
Pearl Culture

A pearl is the result of an injury to the mollusc. Pearl is obtained from pearl oysters. Pearl is a valuable gem known to humankind since ancient times. The origin of pearl is not known to us, however Chinese records show that pearl was known to them as far back as 2300 B. C. pearl is a hard, roundish object produced within the soft tissue specifically the mantle of a living shelled molluscs. Just like the shell of mollusca, a pearl is composed of calcium carbonate in minute crystalline form, which has been deposited in concentric layers. The ideal pearl is perfectly round and smooth, but many other shapes of pearls where obtained. The finest quality natural pearls have been highly valued as gemstones and objects of beauty for many centuries, and because of this, the word pearl became a metaphor for something very rare, very fine, very admirable and very valuable.

The pearl, in fact is of animal origin and produced by certain bivalves of mollusca. The Peral producing bivalves are marine oyesters of the genus Pinctada, through some freshwater bivalves of the genus Unio and Anodonta also produce pearl. Fine gem-quality saltwater and freshwater pearls can and do sometimes occur completely naturally in the wild state, but this is rare. Many hundreds of pearl oysters pearl mussels have to be gathered and opened, and thus killed, in order to find even one wild pearl, and for many centuries that was the only way pearls were obtained. This was the main reason why pearls fetched such extraordinary prices in the past. In modern times however, almost all the pearls for sale were formed with a good deal of expert intervention from human pearl farmers.

Physical Properties

The unique luster of pearls depends upon the reflection, refraction, and diffraction of light from the translucent layers. The thinner and more numerous the layers in the pearl, the finer the luster. The irridescence that pearls display is caused by the

Plate 17

Figure 1: Showing Freshwater *Unio*

Figure 2: Pearls

overlapping of successive layers, which breaks up light falling on the surface. The culture freshwater pearls can be dyed yellow, green, blue, brown, pink, purple, or black.

Natural but cultivated pearl produced by a molluscs after the intentional introduction of a foreign object inside the creature's shell. The discovery that such pearls could be cultivated in freshwater mussels is said to have been made in 13th century China, and the Chinese have been adept for hundreds of years at cultivating pearls by opening the mussel's shell and inserting into it small pellets of mud or tiny bosses of wood, bone, or metal and returning the mussel to its bed for about three years to await the maturation of a pearl formation. Cultured pearls of China have been almost exclusively blister pearls or hemispherical pearls formed between the mussel and its shell, which require addition of a half sphere of mother-of-pearl to create the assembled gem, called a pearl doublet.

The production of whole cultured pearls was perfected by the Japanese. The research that led to the establishment of the industry was started in the 1890s by Mikimoto Kokichi, who, after long experimentation, concluded that a very small mother-of-pearl bead introduced into the mollusc's tissue was the most successful stimulant to pearl production. It possesses the added virtue of providing a pearl entirely of nacreous content. Cultured pearls closely approximate natural pearls. If the covering of nacre is too thin, however, it will deteriorate upon prolonged contact with the acids of the human body and eventually will reveal the mother-of-pearl matrix.

Once a shore-based activity, pearl farms now generally use a vessel as an operating platform. Immature pearl oyster shells usually *Pinctada fucata* or *Pteria penguin* in Japan and *Pinctada maxima* in Australia are reserved in barrels until maturation within 2 to 3 years and, when the shells reach certain size, are implanted with a tiny polished sphere of mother-of-pearl. The implanted oysters are suspended in wire nets from floating rafts or contained in some other way and are tended by divers until they are ready for harvesting; readiness is often determined by X-ray.

Plate 18: Cross-section Showing Pearl Formation

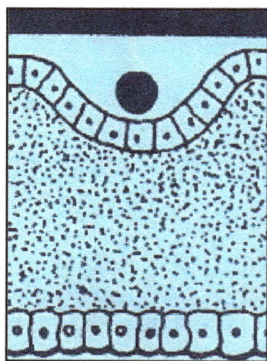

Figure 1:
Containing Foreign

Figure 2: Formation of
Layers of Nagre

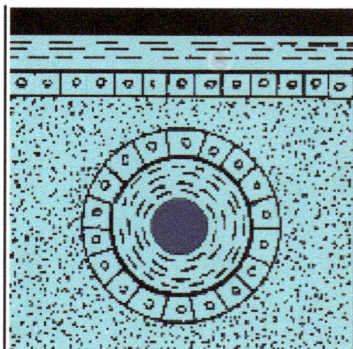

Figure 3: Formation of
Particles Pearl

Pearls are characterized by their translucence and lustre and by a delicate play of surface colour called orient. The shape is spherical or drop like and the deeper its lustre, the greater its value. Only those pearls produced by molluscs whose shells are lined with mother-of-pearl are really fine pearls; pearls from other molluscs are reddish or whitish, porcellaneous, or lacking in pearly lustre. Jewellers commonly refer to saltwater pearls as Oriental pearls and to those produced by freshwater molluscs as freshwater pearls.

The chief component of the nacre that constitutes the pearl is aragonite ($CaCO_3$). Nacre also contains a small amount of conchiolin, a hornlike organic substance i.e albuminoid that is the main constituent of the mollusc's outer shell. The shell-secreting cells of the mollucs are located in the mantle, or epithelium, of its body; when a foreign particle penetrates the mantle, the cells attach to the particle and build up more or less concentric layers of pearl around it. Irregularly shaped pearls called baroque pearls are those that have grown in muscular tissue; pearls that grow adjacent to the shell are often flat on one side and are called blister pearls.

The colour of pearls varies with the mollucs and its environment. It ranges from black to white, with the rose of Indian pearls esteemed most. Other colours are cream, gray, blue, yellow, lavender, green, and mauve. All occur in delicate shades. The surface of a pearl is rough to the touch. Pearls come in a wide range of sizes. Those weighing less than $1/_4$ grain (1 pearl grain = 50 milligrams = $1/_4$ carat) are called seed pearls. The largest naturally occurring pearls are the baroque pearls; one such pearl is known to have weighed 1,860 grains.

The finest Oriental pearls are produced by the *mohar*, a variety of the *Pinctada martensii* species of saltwater mollucs. Found in the Persian Gulf, with the richest harvest taken from the waters off the great bight that curves from the peninsula of Oman to that of Qatar, the pearls come from depths of 8 to 20 fathoms (48 to 120 feet). Other notable sources of fine-quality pearls include the Gulf of Mannar between India and Sri Lanka; the waters off Celebes, Indonesia; and the islands of the South Pacific. In the Americas, the Gulf of California, the Gulf of Mexico, and the waters of the Pacific coast of Mexico have yielded dark-hued pearls with a metallic sheen as well as white pearls of good quality.

Freshwater mussels in the temperate zone of the Northern Hemisphere have produced pearls of great value, as for example those from the Mississippi River. Pearling is a carefully fostered industry in central Europe, and the forest streams of Bavaria, in particular, are the source of choice pearls. Freshwater pearling in China has been known from before 1000 BC. In all pearl fisheries, however, production has declined significantly since the widespread introduction of cultured pearls.

Freshwater and saltwater pearls may sometimes look quite similar, but they come from very different sources. Natural freshwater pearls form in various species of freshwater mussels, family *Unionidae*, which live in lakes, rivers, ponds and other bodies of freshwater. These freshwater pearl mussels occur not only in hotter climates, but also in colder more temperate areas such as in Scotland. The most freshwater cultured pearls sold today come from China.

Saltwater pearls grow within pearl oysters, family *Pterriidae,* which live in oceans. Saltwater pearl oysters are usually cultivated in protected lagoons or volcanic atolls.

Pearl Formation

A pearl is the result of an injury to the mollusc. It is secreted by the mantle as a means of protection against some foreign body. Foreign body is formed between the mantle and the shell as a larva at the stage of fluke. It becomes enclosed in a sac of mantle epithelium which is thus irritated. The irritation stimulates the mantle epithelium to secrete thin connective concentric layers of mother-of-pearl around the foreign body. The amount of deposition is in direct proportion to the degree of irritation. At all the end of several years, a pearl will be found. Pearls are often found in clams an edible oysters but these are not nacreous and therefore of little value. The most precious pearls are found in the pearl oyster which is closely allied to the freshwater mussel. Irritation pearls are beads coated with an irridescent substance called pearl essence that is obtained from the scales of fish and highly valued. Teardrop-shaped pearls are often used in pendants.

Chapter 26

Role of Co-operatives and National Fisheries Development Board in Fisheries

The Co-operative movement in India has gained momentum only after independence. Credit goes to the liberal policies and assistance of Government and various government agencies given to encourage the co-operative sector in the Country. There is a close relation in between Fisheries development and the development of the Fisherman. In India, the socio-economic condition of fisherman is very backward in general. Most of them live below poverty line. The main reason behind this is that this sector is unorganized. Like other industries, Fisheries also need a lot of money for development. The Fisherman should also adopt advanced techniques both in culture and capture fisheries to make this business a profitable venture. Particularly the capture fisheries need a large capital investment for effective exploration of the available resources both inland and marine. Thus, the formation and development of co-operative sector in fisheries is imperative for the development of both, *i.e.* The Fisherman and the Fishery Industry.

Progress of Fisherman Co-operative

Due to special attention paid by the government of India and various state government there is steady progress in fishermen co-operative in India after the independence. Presently the fishermen co-operative is present throughout the country. Their number increased from 200 to 4200 (about) in between 1967 to 2006. The country has 429 Fish Farmers Development Agencies (FFDAs) and 39 Brackishwater Fish Farms Development Agencies (BFDAs) for promoting freshwater and coastal aquaculture.

At national level, Kerala state tanks first in fishermen co-operative societies. There are 817 co-operative societies in this state, out of which 623 are engaged in production while 194 co-operative are the financing bodies. The number of their members gone up to more than 1,03,360. Next is the place of Maharastra, Gujarat and Tamil Nadu. However, by now the number of co-operative societies in different state of our country must have increased considerably, indicating the important role that the co-operative societies are playing.

In inland state also there is substantial progress in fisherman co-operative societies. Almost all states fisheries department have made it a policy to get the fishing done in their reservoirs through fishermen co-operative societies only.

Incentives given to Fisherman Co-operative Societies

Different state government has given various incentives to the co-operative societies of their states as per rules. The major incentives are:

1. Departmental tanks up to 40 ha. are given on lease for 10 years for fish culture.
2. Subsidy up to 50 per cent on procurement of fish seed and inputs.
3. Rights of fishing in reservoirs and large irrigation tanks. The societies get the fish catch just after paying a nominal royalty to state department which is less than 1/3 of the market value of the fish as in case of Madhya Pradesh.
4. Soft term loan are given to the societies with lower interest rates for various developmental purposes.
5. The central agencies like NABARD refinance them on preference.
6. Different kinds of subsidies are given like subsidy up to 25 per cent on Nylone Yarn and up to 33 per cent on cotton yarn for making nest in M.P for the purchase of this yarn loan is also given to these societies as per rules.
7. The fishermen are covered under various insurance schemes like Janata Personal Accident Policy and community Insurance cover for Rs 1,00000 (Rupees one lakh).
8. In maritime states, the fishermen co-operative societies get heavy capital loans for purchase of boats, mechanized boats and other inputs for in-shore and off-shore fishing.

Objectives of National Fisheries Development Board

The following are the main objectives of NFDB:

1. To bring major activities relating to fisheries and aquaculture for focused attention and professional management.
2. To coordinate activities pertaining to fisheries undertaken by different Ministries/Departments in the Central Government and also coordinate with the state/Union Territory Governments.

3. To improve production, processing, storage, transport and marketing of the products of capture and culture fisheries.

4. To achieve sustainable management and conservation of natural aquatic resources including the fish stocks.

5. To apply modern tools of research and development including biotechnology for optimizing production and productivity from fisheries.

6. To provide modern infrastructure mechanism for fisheries and ensure their effective management and optimum utilization.

7. To generate substantial employment.

8. To train and empower women in the fisheries sector.

9. To enhance contribute of fish towards food and nutritional security.

Role of NABARD

National Bank for Agriculture and Rural Development (NABARD) was formed through an act of Parliament (NABARD Act, 1981) by merging Agriculture Refinance Development Corporation (ARDC) and Agriculture Credit Development (ACD) of Reserve bank of India. In order to augment country's fish production and to provide job opportunities to the rural skilled and unskilled youth NABARD's main objective is to provide financial support to all the viable fishery projects. This credit facility is available for the development of Marine, Brackishwater and Inland Fisheries projects. NABARD also associated with the nationalized and Co-operative Banks of our country to support necessary infrastructure so that production of fish, its marketing and export to lucrative markets and better remuneration and returns to fishermen are assured. NABARD mainly covers following area for investment and refinance in Fisheries sectors:

(*a*) Marine Fisheries

1. Purchase of Mechanised Fishing Vessels.

2. Purchase of Non-mechanised Vessels.

3. Establishment of infrastructural facilities like Ice-plants, Cold storage, Canning plants, Processing plants, Net making units, Fish auction halls, Service Stations, Workshop, Banking facilities etc.

4. For Fish Marketing facilities like Carrier Boats, Transport Vehicles, Establishment of Fish Stalls etc.

(*b*) Inland Fisheries

1. Carps and Cat fish farming.

2. Construction of modern Fish Seed hatcheries.

3. Reservoir fisheries Development.

4. Establishment of Fish Seed farms for rearing and producing of fingerlings.

5. Polyculture projects.

6. Cage fisheries

7. Wetland/Beel Fisheries

8. Cold water fisheries.

9. Composite Fish culture

10. Capture Fishery schemes.

(c) Brakishwater Fisheries

1. Monoculture of prawns.

2. Mixed culture of fish-cum-prawn

3. Prawn seed Hatchery projects

4. Larval rearing of prawns

Some Important Role of NFDB for the Benefit of Fishermen

A. Group Accident Insurance Security to Active Fishermen

Fishermen work in some of the most hazardous area prone to accidents. Presently the help of NFDB some State Government provides insurance coverage to 50,000 active fishermen under National Scheme for Welfare of Fishermen. It is proposed to provide insurance coverage to more active fishermen. Apart from this active fishermen would be provided coverage under Janshree Bima Yojna. Under this scheme Rs. 50.00 per person would be provided to the insurance company and Rs. 50.00 would be given by the beneficiaries himself. The rest Rs. 100.00 would be provided by the Social Security Fund. Twenty thousand fishermen are being targeted to be insured during 2008-09.

B. Model Fishermen Village

Houses, safe drinking water and community hall are given to poor fishermen under National Scheme of Welfare of Fishermen, Unit cost of a house is Rs. 40,000 and that of a hand pump is Rs. 12,000. One community hall is constructed where more than 75 houses are built. Houses are built by the fishermen themselves and the amount for this is given to them by cheque. It is a centrally sponsored scheme in which is cost is shared by the State government and GOI on 50:50 basis.

C. Strengthening of FFD

There are 33 FFDAs operational in the State. Presently their financial activity is very limited and they are not working as independent organistaion. To make them effective and for operationally strengthening them it is proposed to provide them a working capital of Rs. 10.00 lakhs. This will provide speedy implementation of developmental schemes. With this FFDAs will manage their own business. This fund would be used as a revolving fund.

D. Crop Insurance Scheme

It is being proposed to bring fish crop in a pond under insurance coverage. With this farmers will get insurance benefit in case of their crop damage by natural calamity or damage by miscreants. The premium for the crop insurance may be shared by beneficiaries and the State Govt. A detail scheme is under consideration for the same. The present calculation is based upon a scheme by the Oriental Insurance Company in which cost of fish (per hectare production) has been kept at Rs. 40,000. Half of the premium would be borne by the farmer or the settlee of the pond.

E. Renovation of Ponds

Fisheries ponds in the State are very old. Due to frequent floods along with natural weathering and erosion they have lost their optimum carrying capacity. Though efforts are being made through schemes like RSVY and National Rojgar Guarantee Yojna, a targeted approach is necessary. Similarly subsidy based instruction of new ponds needs the similar approach.

F. Matsya Krishak Samman Yojna

With the aim to encourage fish farmers and promoting them to increase productivity Matsya Krishak Samman Yojna is being proposed. Farmers would be given Fisheries and Aquaculture instruments *i.e.*, aerator, pumping set, water and soil testing kit etc. in Samman Yojna. Those fish farmers who have attained an annual production level of 3000 kg fish/ha may qualify for this scheme. Cumulative financial estimate for four years is Rs. 35.00 lakh.

References

Alikunhi, K. H. (1957): Fish culture in India. *Fm. Bull. Indian Coun. Agri. Res.20: 144p.*

Alikunhi, K. H; Sukumaran, K. K; Parmeswaran, S and Banerjee, S.C (1964): Preliminary observation on commercial breeding of Indian Carp under controlled temperatures in the laboratory. *Bull. Cent. Inl. Fish Res. Inst. Barrackpore* 3: 20p.

Anonymous (2001): CPF's meatier strain of *Macrobrachium rosenbergii*. *Asian Aquaculture Magazine*, September/October 2001:10-11.

Beamish, F.W.H. (1964): Seasonal changes in the standard rate of oxygen consumption of fishes. Can. J. Zool., 42:189-194.

Bennett. A.F. (1988): Structural and functional determinates of metabolic rate. Amer. Zool., 28:699-708.

Bove, J. (1962): MS$_{222}$ Sandoz: *The anaesthetic of choice for cold blooded organisms.* Sandoz News No 3, 12.

Chakrabarty, R.D. and Murty, D.S. (1972): Life history of Indian major carps *Cirrhinus mrigala* (Ham.), *Catla catla* (Ham.) and *Labeo rohita* (Ham.). J. Inland Fish. Soc. India. 4:132-161.

Chandar, S. L (1970): *Hypophysation of Indian major Carps*, 100p. Satish Enterprises, Agra.

Chaudhuri, H. (1963): Induced spawning of Indian carps. *Proc. Nat. Sci. India*, 29B (4): 478-87.

Chaudhuri, H. and Alikunhi, K. H. (1957): Observations on the spawning in Indian carps by hormone injection. *Curr. Sci.* 26(12): 381-82.

Chaudhuri, H., Singh, S. B., Sukumaran, K. K and Chakarbarti, P. C. (1967): Note on natural spawning of grass carp and silver carp in induced breeding experiments. *Sci. Cult.*33: 493-94.

Chowdhury, R., Bhattacharjee, H. and Angell, C. (1993): *A manual for operating a small-scale recirculation freshwater prawn hatchery.* Publication No. BOBP/MAG/13. Madras, India, Bay of Bengal Programme.

CIFE (1990): Management of freshwater fish seed farms, 120p.

Colesante, R.J. (1977*). Improvements of osocid culture techniques.* New York Department of Environment of Conservation, Report of 1976 Studies, Albany.

Daniels, W.H., D'Abramo, L.R., Fondren, M.W. and Durant, M.D. (1995): Effects of stocking density and feed on pond production characteristics and revenue of harvested freshwater prawns *Macrobrachium rosenbergii* stocked as size-graded juveniles. *Journal of the World Aquaculture Society,* 26:38-47.

Das, P.C; Ayyappan, S and Jena, J.K. (2003): Oxygen consumption pattern of fingerlings of Indian major carps, *Catla catla* (Ham.), *Labeo rohita* (Ham.) and *Cirrhinus mrigala* (Ham.). J. Inland Fish. Soc. India, 35(2):45-49.

Datta, H. M and Datta Munshi, J. S (1985): Functional morphology of air breathing fishes: A review. *Pro. India Acad. Sci.,* 94(4): 359-375.

Desilva, C.D. and Tytler, P. (1973): The influence of reduced environmental oxygen on the metabolism and survival of herring and plaice larvae. Neth. J. Sea Res., 7:345-362.

Dubey, G. P and Tuli, R. P (1961): Observations on the breedings of the major carps in Madhya Pradesh. *J. Bombay nat. Hist. Soc.* 58(1): 81-91.

FAO (1981): *Water for freshwater fish culture.* FAO Training Series No. 4. Rome.

FAO (1985): *Soil and freshwater fish culture.* FAO Training Series No. 6. Rome.

FAO (1988): *Topography for freshwater fish culture: topographical tools.* FAO Training Series No. 16/1. Rome.

FAO (1989a): *Aquaculture production (1984-1986).* FAO Fisheries Circular No. 815. Rome.

FAO (1989b): *Topography for freshwater fish culture: topographical surveys.* FAO Training Series No. 16/2. Rome.

FAO (1992a): *Simple economics and bookkeeping for fish farmers.* FAO Training Series No. 19. Rome.

FAO (1992b) *Pond construction for freshwater fish culture: pond farm structures and layouts.* FAO Training Series No. 20/2. Rome.

FAO (1994): *Handbook on small-scale freshwater fish farming.* FAO Training Series No. 24. Rome.

FAO (1995): *Pond construction for freshwater fish culture: building earthen ponds.* FAO Training Series No. 20/1. Rome.

FAO (1996): *Management for freshwater fish culture: ponds and water practices.* FAO Training Series No. 21/1. Rome.

FAO (1998): *Management for freshwater fish culture: fish stocks and farm management.* FAO Training Series No. 21/2. Rome.

FAO (2001): *Fishery statistics: aquaculture production (1999).* FAO Fisheries Series No. 58/FAO Statistics Series No. 160. Rome.

FAO (2002): *Fishstat Plus (v.2.30),* 15.03.2002. Rome, FAO.

FAO (2006): *Prospective Analysis of Future Aquaculture Development and the role of COFI sub-committee on Aquaculture.* Third session, committee on Fisheries: Sub-Committee on Aquaculture. New Delhi, India, 4-8 September, 2006. COFI: AQ/ III/2006/8. Food and Agriculture Organization, Rome.

FAO (2007): *State of world Fisheries and Aquaculture,* 2006. Food and Agriculture Organization, Rome.

Fred Ward, (2002): *Pearls (Fred Ward Gem Book), 3rd Edition,* Gem Guides Book Company, pgs. 35-36.

Fry, F. E. J (1971): The effect of environmental factors on the physiology of fish. In Fish Physiology (Eds: W. S. Hoar and D.J. Randhall) Accademic Press, New York, pp. 1-98.

Ghosh, T. K., Moitra, A. and Kunwar, G.K. (1986). Bimodal oxygen uptake in a freshwater air-breathing fish, *Notopterus chitala. Jap. Jour. Ichthyol.* 33 (3): 280-285.

Ganpati, S. V. and Chaco, P. I. (1950): A Comperative study of the transport of fish seed in oxygen tin carriers and in ordinary tin carriers. *Ind. Com. J.* 6: 564-87.

Ghosh, T.K. and Munshi, J.S.D. (1987): Bimodal oxygen uptake in relation to body weight and seasonal temperature of an air-breathing climbing perch, *Anabas testudineus* (Bloch). Zool. Beitr. N.F., 31(3):357-364.

Ghosh, T.K.; Kunwar, G.K. and Munshi, J.S.D. (1990): Diurnal variation in the bimodal oxygen uptake in an air-breathing catfish, *Clarias batrachus.* Japan. J. Ichthyol., 37(1):56-59.

Ghosh, T. K and Jha, K. K (1994): Packing and long distance transportation of certain live air-breathing teleosts from India. *American Fish Soc.124th Ann. Mett. Halifax, Nova, Scotia,* 21-25 Aug.pp.91.

Ghosh, T. K, Jha, K. K and Roy, P. K (1994): Effect of an Anaesthetic on the oxygen uptake in the juvenils of an Indian major carp *Catla catla* (Ham). *Proc. 81st Indian Sci. Cong.Jaipur Sec.Zool., Entomol. and Fish.* pp. 57.

Giguere, L.A.; Cote, B. and St-Pierre, J.F. (1988): Metabolic rates scale isometrically in larval fishes. Mar. Ecol. Prog. Ser., 50:13-19.

Haltingh, G. L. and Burger, A.P. (1979). Haematological assessment of the effects of the anaesthetic MS_{222} in natural and neutralized form in three freshwater fish species: intra-species differences. *J. Fish. Biol.* 15, 645-653.

Hora, S. L. (1945): Analysis of factors influencing the spawning of major carps. *Proc. Nat. Inst. Sci. India.*11(3): 303-12.

Hora, S. L. and Pillay, T. V. R. (1962): Handbook on fish culture in the Indo-Pacific region. *FAO Fish. Biol. Tech. pap.* (14): 204p.

Houstin, A. H., Madden, J. A., Woods, R.J. and Miles, H. M. (1971). Some physiological effects of handling and tricaine methane sulphonate anaesthetization upon the brook trout *(Salvelinus fontinalis). J. Fish. Res. Bd. Can.* 28: 262-633.

IIMA(1983): National issues raised by IIM: *Study on development and management of freshwater aquaculture in India*-A review by V.R.P. Sinha and K.K. Ghosh, CIFE, Bombay.

Imabayashi, H and Takahashi, M (1987): Oxygen consumption of post larval and juvenile Red Sea Bream, Pagrus major, with special reference to group effect. *J. Fac. Appl. Biol. Sci.* 26:15-21.

Ismael, D. and New, M.B. (2000): Biology. *In* M.B. New and W.C. Valenti, eds. *Freshwater prawn culture: the farming of Macrobrachium rosenbergii,* pp. 18-40. Oxford, England, Blackwell Science.

Jha, K. K (1993): Effect of water from different sources on the survival of hatchlings of certain Teleosts. *J. Freshwater Biol. 5(2): 153-157.*

Jha, K. K (2007): Jawv Vivdata Arunachal Rajya Ke Adi Janjati dwara Matyakhate Ki Paramparik Mathshyan Vibidhiya–Disaya Wa Aayam. P 23–29, K. K. Vas, N. P. Srivastva, P. K. Katiya, Krishna Mitra, P. R. Rao (Eds.), Matashyaki Anusandhan Awang Vikash–Disaya Wa Aayam, *Central Inland Fisheries Research Institute, Barrackpore,* Kolkata.

Jha, K. K (2008): Ramro Korang Lake needs studies on fish diversity and water quality Analysis for aquaculture. *Curr. Sci.,* 95(9): 1107–1108.

Jha, K. K, Pandit, D.N and Ghosh, T. K (2004): A comparative study of aquatic oxygen uptake under sedation in the early developmental stages of *Catla catla* (Ham) and *Labeo rohita* (Ham.).*J. Inland Fish. Soc. India 26(2): 61-64.*

Jha, K. K., Roy, P. K and Ghosh, T. K (2008): Oxygen uptake in the early developmental stages of two Indian Major Carps, *Catla catla* (Ham.) and *Labeo rohita* (Ham.) in relation to body weight. *J. Inland Fish. Soc.India,* 40(2): 7-12.

Jhingran, V. G (1968): Riverine fish seed resources and their exploitation in India. In: *Seminar on production of Quality Fish Seed for Fish Culture.* Sponsered by ICAR at CIFRI, Barrackpore from November1-2,1968.

Jhingran, V. G (1991): *Fish and Fisheries of India,* 727p. Hindustan Publishing Corporation (India) Delhi.

Kamler, E. (1976): Variability of respiration and body composition during early developmental stages of carp. Pol. Arch. Hydrobiol., 23(3):431-485.

Kewalramani, H. G. and Gogate, M. G. (1968). Anaesthetisation in fish, *Tilapia* and major carps. *Proc. Indian Acad. Sci.,* 67 B (50): 237-246.

Khan, H. (1924): Obsevations on the breeding habits of some freshwater fishes in the Punjab. *J. Bombay nat. Hist. Soc.* 29(4): 958-62.

Khan, H. (1938): Ovulation in fish (effect of administration of anterior lobe of pituitary gland). *Curr. Sci.* 7(5): 233-34.

Khan, H. (1945): Observations on the spawning behaviour of carp in Punjab. *Proc. Nat. Inst. Sci. Ind.* 11(3): 315-20.

Kidd, R.B. and Banks, G.D.(1990). Anaesthetizing lake trout with Tricaine (MS_{222}) administered from a spray bottle. *The Prog. Fish-cult.* 52: 272-273.

Kleiber, M. (1961): *The fire of life: An Introduction to animal energetics.* Wiley and Sons, New York.

Kunwar, G.K.; Pandey, A. and Munshi, J.S.D. (1989): Oxygen uptake in relation to body weight of two freshwater major carps *Catla catla* (Ham.) and *Labeo rohita* (Ham.). Indian J.Ani. Sci., 59(5):621-624.

Lakshmanan, M.A.V., Murty, D.S., Pillai, K. K. and Banerjee, S.C (1967): On a new artificial feed for carp fry. *FAO Fish. Rep* (44)3:373-87.

Laurence, G.C. (1978): Comparative growth, respiration and delayed feeding abilities of larval cod (*Gadus morhua*) and haddock (*Melanogrammus aeglefinus*) as influenced by temperature during laboratory studies. Mar. Biol., 50:1-7.

Lavens, P. and Sorgeloos, P. (eds. 1996): Manual on the production and use of live food for aquaculture. FAO Fisheries Technical Paper No. 361. Rome.

Mc Farland, W. N. (1959). A study of the effects of anaesthetics on the behaviour and physiology of fishes. *Publ. Inst.Mar. Sci.*, 6: 23-55.

Mc Farland, W.N. (1960). The use of anaesthetics for handling and transport of fish. *Calif. Fish. Game.*, 66(4): 408-431.

Meyer, F.P., Schnick, R.A., Cumming, K. B. and Berger, B.L. (1976). Registration status of fishery chemicals. *Prog. Fish.-cult.*, 38(1):3-7.

Mishra, N., Ojha, J. and Munshi, J.S.D. (1984): Haematology of juveniles and adults *Channa marulius*. Arch. Hydrobiol. 100(2):267-272.

Moss, D.D. and Scott, D.C. (1961): Dissolved oxygen requirements of three species of fish. Trans. Amer. Fish. Soc., 90:377-393.

Motwani, M.P. and Bose, B.B. (1957): Oxygen requirements of fry of the Indian major carp *Labeo rohita* (Ham.). Proc. Nat. Inst. India, 23B(1-2):8-16.

Munshi, J.S.D. and Dube, S.C. (1973): Oxygen uptake capacity of gills in relation to body size of the air-breathing fish *Anabas testudineus* (Bl). Acta. Physiol., 44:113-124.

Munshi, J.S.D.; Pandey, B.N.; Pandey, P.K. and Ojha, J. (1978): Oxygen uptake through gills and skin in relation to body weight of an air-breathing siluroid fish, *Saccobranchus* (= *Heteropneustes*) *fossilis*. J.Zool. Lond., 184: 171-180.

Munshi, J.S.D.; Patra, A.K. and Hughes, G.M. (1982): Oxygen consumption from air- and water in *Heteropneustes* (= *Saccobranchus*) *fossilis* (Bloch) in relation to body weight at three different seasons. Proc. Indian Natn. Sci. Acad., B 48(6):715-729.

Neil H. Landman, *et al.* (2001) *Pearls: A Natural History*, Harry Abrams, Inc., 232 p.,

New, M.B. (1995): Status of freshwater prawn farming: a review. *Aquaculture Research*, 26:1-54.

Nieminen, M., Laitinen, M. and Pasanen, P. (1982). Effects of anaesthesia with Tricaine (MS_{222}) on the blood composition of the splake (*Salvenlinus fontinalis* x *Salvelinus namaycush*) *Comp. Biochem. Physiol.*, 73C: 271-176.

Odum, E. P.(1971): *Fundamentals of Ecology*. 3rd edn., W. B. Saunders, New York.

Oikawa, S. and Itazawa, Y. (1985): Gill and body surface areas of the carp in relation to body mass with special reference to the metabolism- size relationship. J. Exp. Biol., 117: 1-14.

Ojha, J., Dandotia, O.P. and Munshi, J.S.D. (1977): Oxygen consumption of an amphibious fish, *Colisa fasciatus* in relation to body weight. Pol. Arch. Hydrobiol., 24(4):547-553.

Ojha, J., Dandotia, O.P., Patra, A.K. and Munshi, J.S.D. (1978): Biomodal oxygen uptake in relation to body weight in a freshwater murrel, *Channa* (= *Ophiocephalus*) *gachua*. Z. Tier-physiol., 40: 57-66.

Pandit, D.N. (2001): Studies on the oxygen requirement of early developmental stages of certain air-breathing teleosts in relation to their transportation. Ph.D. thesis, T.M. Bhagalpur University, Bhagalpur, 83 Pp.

Pandey, B.N.; Pandey, P.K.; Choubey, B.J.P. and Munshi, J.S.D. (1976): Studies on blood components of an air-breathing siluroid fish *Heteropneustes fossilis* in relation to body weight. Folia Haematol. 103(1): 101-116.

Pearl oyster farming and pearl culture http://www.fao.org/docrep/field. The origin of the Japanese akoya.

Prigogine, I and Wiame, J.M. (1946): Biologie el thermodynamique des phenomenes irreversibles. Experiential, 2: 451-453.

Reddy, G.A. and Rao, P.L.M.K. (2001): Freshwater prawn farming: a proven success in India. *Fish Farmer*, 24(5):32-34.

Roy, P.K. and Munshi, J.S.D. (1984): Oxygen uptake in relation to body weight and respiratory surface area in *Cirrhinus mrigala* (Ham.) at two different seasonal temperatures. Proc. Indian natn. Sci. Acad., B50(4): 387-394.

Saha, G.N. (1966): A note on the oxygen requirement of spawn of Indian major carps. Sci. Cult., 32: 327-328.

Saha, K.C., Sen, D. P., Roy Choudhary, A. K. and Chakravarty, S. K. (1957): Studies on factors influencing spawning of Indian major carps in "bundh" fisheries. *Indian J. Fish*.4(2): 284-94.

Schmidt-Nielsen, K. (1984): *Scaling; Why is animal size so important?* Cambridge University Press, Cambridge.

Sundararja, B. I. and Goswami, S. V. (1974): Comperative aspect of oocyte maturation in sub-mammalian vertebrates. A review. *Proc. Indian nat. Sci. Acad.* 33B(3): 286-95.

Schoettger, R.A. and Julin, A.M. (1967). Effect of MS$_{222}$ as an anaesthetic on four Salmonies 13: pp 15.

Shlaifer, A (1938): Studies in mass Physiology: Effect of numbers upon the oxygen consumption and locomotor activity of *Carassius auratus. Physiol. Zool.* 11: 408–424.

Singh, B.N. (1977): Oxygen consumption and the amount of oxygen required for transport of *Rohu* and *Mrigal* fingerlings. J. Inland Fish. Soc. India, 9: 98-104.

Singh, O. N and Datta munshi, J. S (1985): Oxygen uptake in relation to body weight, respiratory surface area and group size in freshwater goby, *Glossogobius giuris* (Ham.). *Proc. Ind. Nat. Sci. Acad.* B51(1): 33–40.

Singh, D.K.; Singh, O.N. and Munshi, J.S.D. (1991): Oxygen uptake in relation to body weight in *Rita rita* (Ham.) (Bagridae, Pisces) at two different seasonal temperatures. J. Freshwater

Smit, G.L., Haltingh, J. and Burger, A.P. (1979). Haematological assessment of the effects of the anaesthetic MS$_{222}$ in natural and neutralized form in three freshwater fish species: intraspecies differences. *Fish. Biol.,* 15: 645-653.

Soivio, A., Nayholm, K. and Huhti, M. (1977). Effects of anaesthesia with MS$_{222}$, neutralized MS$_{222}$ and Benzocaine on the blood constituents of rainbow trout, *Salmo gairdneri. J. Fish. Biol.,* 10: 91-101.

Tave, D. (1999): *Inbreeding and broodstock management.* FAO Fisheries Technical Paper No. 392. Rome.

Waresh, W. D and Igram, R (1978): Oxygen consumption in the fathead minnow (*Plmephales promelas Rafinesque*)–1. Effect of weight, temperature, group size, oxygen level and opercular movement rate as a function of temperature. *Comp. Biol. Chem.Physiol.* 62A: 351-356.

Wedemeyer, G. (1970). Stress of anaesthesia with MS$_{222}$ and benzocaine in rainbow trout (*Salmo gairdneri*). *J. Fish. Res. Bd. Can.* 27: 909-914.

Welch, P. S (1948): *Limnological Methods.* McGraw Hill, New York.

Wheaton, F.W. (1977): *Aquacultural engineering.* New York, John Wiley and Sons.

Index

www.ingramcontent.com/pod-product-compliance
Lightning Source LLC
Chambersburg PA
CBHW050518190326
41458CB00005B/1581